POLLUTION CONTROL IN THE DAIRY INDUSTRY

POLLUTION CONTROL IN THE DAIRY INDUSTRY

Harold R. Jones

NOYES DATA CORPORATION

Park Ridge, New Jersey   London, England

1974

Published in the United States of America by
Noyes Data Corporation
Noyes Building, Park Ridge, New Jersey 07656

FOREWORD

This Pollution Technology Review is based on authoritative government reports. It attempts to clarify the ways and means open to the alert dairy processor who must keep his polluting wastes down to a minimum.

Most effluent wastes from the dairy industry are biodegradable, but treatment costs are increasing, effluent discharge requirements are becoming more stringent, and urbanization increasingly limits the availability of land. Thus, there are many problems to be dealt with in handling the industry's waste. This is especially true in the case of whey disposal.

In the United States, we are fortunate in receiving direct help from the numerous surveys, together with active research and development programs that are being supported by the Federal Government to help industry control its wastes and troublesome effluents.

In this book are condensed vital data from government sources of information that are scattered and difficult to pull together. Important processes are interpreted and explained by actual case histories. This condensed information will enable you to establish a sound background for action towards combating pollution in the dairy industry.

Advanced composition and production methods developed by Noyes Data are employed to bring our new durably bound books to you in a minimum of time. Special techniques are used to close the gap between "manuscript" and "completed book." Industrial technology is progressing so rapidly that time-honored, conventional typesetting, binding and shipping methods are no longer suitable. We have bypassed the delays in the conventional book publishing cycle and provide the user with an effective and convenient means of reviewing up-to-date information in depth.

The Table of Contents is organized in such a way as to serve as a subject index and provides easy access to the information contained in this book.

CONTENTS AND SUBJECT INDEX

INTRODUCTION

The U.S. dairy industry had 245,000 employees with a payroll of over $1 billion in 1970 (1). The gross national product is enriched by $12.9 billion from the dairy industry. The dairy products industry accounts for 16% of the value of all food shipments in the United States.

In reviewing trends in the dairy industry, Harper et al (2) found that the average fluid milk plant processes about 100,000 pounds per day. By 1983 they predicted the average plant would produce 225,000 pounds of fluid milk per day. Current trends suggest a significant ecological impact of the industry in that the biggest pollution contributors in the dairy industry are the fastest growing, i.e., cottage cheese, ice cream and cheese. Whey is widely recognized as the largest wasteload not economically subject to use in by-products (3)(4)(5)(6). Figure 1 was adapted from Harper et al (2) to show the trends of production of dairy products.

The dairy food industry is an important part of the overall food industry and contributes materially to fluid wastes which must receive treatment prior to their discharge to streams and other waterways.

Industrial wastes from dairy plants consist primarily of varying quantities of waterborne milk solids from a variety of sources, detergents, sanitizers and lubricants and domestic wastes. The quantity and strength of the wastewater discharged from the dairy food plants vary widely depending upon the quantity of water utilized, the type of process in the operation, and the control management exerts over various waste discharges.

Jones pointed out that dairy plant operators too often take water for granted (7). He suggested that the first step in reducing water costs is to analyze each operation for water use.

FIGURE 1:   PAST AND PROJECTED ANNUAL PRODUCTION OF
SELECTED DAIRY FOODS IN THE UNITED STATES

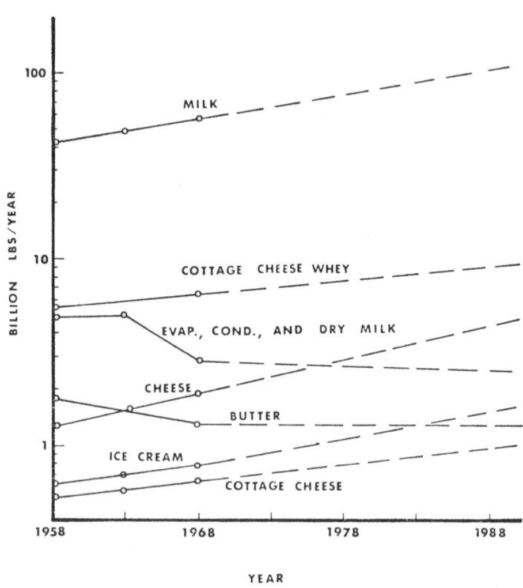

Source: Report PB 220,704

Also, positive steps to help conserve water were outlined. Zall and
Jordan (8) concluded that a water management program in a small dairy
had reduced water use and effluent BOD by 50%. They also presented
the costs of water and the cost involved in letting hot effluent into the
sewer with no attempt to recover the thermal energy. Watson (9) pointed
out the significance of dairy plant effluents and necessity of control of
plant operations to prevent product losses.

Watson (10) has also outlined the dairy processing industry and explained
the sources of waste. A review of dairy waste treatment plants also was
presented. The Public Health Service prepared a manual for the dairy industry detailing the sources and volumes of wastes, effects of dairy pollution, waste prevention and a bibliography on milk wastes (11). Nemerow
also reported on dairy wastes including the average composition of milk
products which might be wasted and an extensive bibliography of 153 publications (12).

Harper et al (2) completed in 1971 a state-of-the art investigation of the
United States dairy industry. They have extensively reviewed the litera-
ture on water and wastewater control and treatment in the dairy industry.
This work is one of various publications applicable to the modern dairy

and is highly recommended for detailed study.

Others are The Treatment of Dairy Plant Wastes (13) prepared by K.S.
Watson of Kraftco Corp. for the Environmental Protection Agency's Technology Transfer Program and Water and Waste Management in Dairy Processing by Carawan, Jones and Hansen (14). The Harper et al publication (2) cited above parallels another report by Harper and Blaisdell (15).

The most visible pollutant in dairy food plant wastewaters is whey from
cheese and cottage cheese operations with about 17 billion pounds of
sweet whey being produced annually in this country from ripened cheese
and about 6 billion pounds of acid whey coming from cottage cheese and
allied processes. The next most significant source of BOD_5 is derived from
every 1,000 pounds of milk processed into cottage cheese from whey and
another 8 pounds comes from the wash water. Under average conditions,
the total BOD_5 from all other dairy food plant processes combined is about
2 to 5 lbs./1,000 lbs. of milk processed.

Little research in respect to dairy food plant wastes and waste treatment
has been conducted in the U.S. in the past decade, except that which has
centered around the visible whey disposal problems, according to Harper
et al (15).

Generally, the dairy food industry knows the problem of fluid waste and
waste treatment will have to be faced, but has not taken the initiative in
solving the problem and with a few exceptions it appears that the industry
will not take remedial steps until forced to do so by legal and/or economic
pressures.

A major reason for the passiveness of the industry is economic. Because
of low profit margins, the expenditure of funds for noneconomic terms has
been avoided. The concept that the dairy food industry must consider water
pollution control as an integral cost of doing business, like any other production item, has not been accepted. In addition to economic, other
reasons for industry's passive attitude are (a) a broad lack of knowledge of
the nature and strength of dairy food plant waste, (b) a failure to under-
stand the potential economic value in recovering wastes and converting
them to usable by-products, and (c) a general lack of understanding of the
technology of waste control and treatment. Numerous members of the industry who are not currently involved with operating waste treatment facilities or paying surcharges on BOD_5 or other waste composition components,
have little understanding of BOD, and most have never heard of COD.

In the course of a study of dairy food plant wastes and waste treatment practices conducted by the Department of Dairy Technology of Ohio State University under Grant No. 12060-EGU from the Water Quality Office of the

Environmental Protection Agency (3), a survey was made of all major
proprietary and cooperative dairy food firms in the U.S., and information
was obtained on current knowledge of about 697 plants in 38 states. These
plants comprise about 11% of the total dairy food plants in the country,
but process more than 65% of the total milk supply.

In addition, plant visitations were made to 30 dairy food plants. A major
emphasis was placed on the large, modern and automated plant, since
these types of operations are becoming the dominant type of dairy food
plant operation, and these large operations can be expected to continue
to increase. The significance of these plants can be illustrated by the fact
that 10 of the plants visited process about 2% of all the milk produced in
the U.S. at the present time.

Contacts were made with the management of the 18 largest proprietary
dairy companies and the 10 largest farm cooperatives processing dairy
foods to ascertain their knowledge of the wasteloads of the plants which
they operated. Attention was centered on knowledge of BOD in plant
waste discharges as an index of knowledge of wasteloads. Information was
also obtained concerning the use of waste composition as a basis for sewage
surcharges and information concerning dairy plant treatment of their own
wastes to determine the relationship of these factors to knowledge of wasteloads.

Information was obtained from all of the proprietary organizations and all
but three of the cooperatives. In addition, the same information was obtained for 234 independent dairy plants in Alabama, Illinois, Indiana,
Iowa, Kentucky, Massachusetts, Michigan, Missouri, New York, Ohio,
Pennsylvania, Tennessee and Wisconsin. In all, 647 plants were included
in the survey (about 10% of the total dairy plants in the country). Although precise figures cannot be obtained, the plants represented processed
more than 65% of all the milk sold off the farm in 1969.

The information in respect to the knowledge of wasteloads, number of
plants paying surcharge on either BOD or suspended solids, the number
of plants which either have their own treatment facility, and the number
of plants discharging untreated wastes to municipal sewer systems, is summarized in Table 1. The information is divided into three categories:
national or regional proprietary companies, local proprietary companies
and major dairy cooperatives operating processing plants. Overall, the
knowledge of BOD concentration in wastewater was about 11%. Statistical analysis, on the basis of the available data, shows that the knowledge
of wasteloads (for the total industry) could not exceed 20% at a 1% confidence limit. Thus, it would appear that the knowledge of wastewater
BOD loads in the dairy food industry is small and is insufficient to provide
a true index of the total BOD and SS from the nation's dairy food industry

TABLE 1:   SUMMARY OF KNOWLEDGE OF DAIRY FOOD PLANT LOADS BY DAIRY FOOD FIRMS

Proprietary Companies	No. of Plants	Knowledge of Waste Loads		Surcharge on Waste Composition		Plants Treating Own Waste		Discharge to City	
		No.	%	No.	%	No.	%	No.	%
A	32	0	0	0	0	7	21.8	25	78.1
B	35	2	5.7	2	5.7	1	2.8	35	100
C	9	4	44.4	2	22.2	0	0	8	88.8
D	19	3	15.7	2	10.5	1	5.2	18	---
E	75	3	4	2	2.6	1	1.3	74	98.6
F	10	2	20	0	0	2	20	7	70
G	15	2	13.3	1	6.6	1	6.6	13	86.6
H	12	1	8.3	1	8.3	0	0	12	100
I*	5	5	100	0	0	1	20	4	80
J	11	0	0	0	0	0	0	11	100
K	8	0	0	0	0	0	0	8	100
L	27	6	22.2	1	3.7	6	22.2	22	81.4
M	20	2	10	0	0	3	15	18	90
N**	7	6	---	---	---	3	---	4	---
O	15	2	2	2	13.3	1	6.6	14	93.3
P	5	2	40	2	40	0	0	5	100
Q	15	1	6.6	1	6.6	2	13.3	14	93.3
Local Proprietary Companies	234	28	11.9	20	8.5	16	6.8	224	95.7
Cooperatives									
R	9	1	11.1	0	0	4	44.4	---	---
S	5	0	0	0	0	0	0	---	---
T	1	1	100	0	0	1	100	---	---
U	27	1	3.7	5	18.5	3	11.1	---	---
V	21	0	0	0	0	0	0	21	100
W	7	0	0	0	0	0	0	7	100
X	23	3	13	3	13	2	8.6	20	86.9
Total	647	75	11.5	44	6.8	55	8.6	564	87.1

(Rows A through Q grouped as National and Regional.)

*Recently closed two plants after municipalities imposed surcharge on BOD and suspended solids.

**Incomplete information on selected operations; percentages not calculated.

Source: EPA Report 12060 EGU, March 1971

that requires waste treatment at this time. In six instances where initial surveys for BOD and SS were made while this study was in progress, the values obtained were 1.5 to 10.0 times the values for other plants in the same organization for which BOD and SS values had been known for some time.

At the same time, caution must be employed in interpreting the extent of wasteload knowledge for various plants. Frequently, the information available is on the basis of a single survey and only a limited number of plants have information of wasteloads on a continuing basis. The industry as a whole appears to be in a very vulnerable position, and no firm statement on the composition of dairy wastes can be made on an industry-wide basis at this time.

Companies differ widely in respect to their knowledge of wasteloads of their various plants. Companies operating more than one processing plant had knowledge of BOD values ranging from 0 to 100%. The organization having 100% knowledge was an exception, having as a key administrator an individual with research experience in dairy wastes. Generally, the proprietary companies are better informed than the cooperatives about their waste composition (12.8 to 6.4%, respectively).

The regional supermarket companies that have become a market force in the past decade generally have more knowledge of wasteloads than more established firms because they have built their plants at a time when states and communities were more waste conscious. Generally, where plant age was known, knowledge of wasteloads was about four times greater in plants built since 1960 than for plants built previously. The major factor would appear to be that these are larger plants in small cities and suburbs with a much more visible wasteload.

Overall, 87% of the plants discharged their wastewaters to municipal treatment plants and 9% treated their own wastes. About 80% of the plants paid a surcharge on wastewater volume, whereas only 7% paid a surcharge on BOD or other waste composition factors.

The breakdown of knowledge of wasteloads by various segments of the dairy food industry is presented in Table 2. Information concerning categorization of plants by the types of products manufactured was available for 445 of the 647 plants surveyed. 404 of these plants discharged wastes to municipal treatment plants, 34 treated their own wastes, 43 paid a surcharge on BOD levels and 72 had knowledge of the BOD_5 in their wastewaters. 174 (39%) of the plants processed more than one type of dairy product in the facility.

TABLE 2:   SUMMARY OF DAIRY FOOD PLANT WASTE KNOWLEDGE BY PLANT TYPE

Type of Operation	No. of Plants	No. of Plants Discharging Waste to-City	No. of Plants With Knowledge of BOD Loading	No. of Plants With Treatment Facilities	No. of Plants Paying Sur-Charge on BOD
Milk	143	138	14	5	13
Ice Cream	26	26	2	0	2
Cottage Cheese	11	8	3	2	0
Conc. Milks	22	13	10	9	0
Milk Powder	16	12	2	4	4
Ripened Cheese	53	40	18	13	0
Milk and Ice Cream	21	21	5	1	6
Milk and Cottage Cheese	43	42	8	2	9
Milk, Ice Cream and Cottage Cheese	51	49	5	1	5
Milk and Other Products	59	55	5	4	4
Total	445	404	72	41	43

Source: EPA Report 12060 EGU, March 1971

TABLE 3: PERCENT KNOWLEDGE OF BOD

Type of Operation	No. of Plants Discharging Waste to City, %	No. of Plants With Knowledge of BOD Loading, %	No. of Plants With Treatment Facilities, %	No. of Plants Paying Surcharge on BOD, %
Milk	96	9.8	3	9
Ice Cream	100	8.3	--	8.3
Cottage Cheese	75	12.5	12.5	--
Conc. Milks	59	4.5	40.9	--
Milk Powder	75	12.5	25	25
Cheese	75.0	24.4	22.4	--
Milk and Ice Cream	100	25	5	30
Milk and Cottage Cheese	97.6	16.6	4.7	19
Milk, Ice Cream and Cottage Cheese	96	9.8	1.9	9.8
Milk and Other Products	93.2	8.4	6.7	6.7
Total	90.8	15.9	9.0	9.8

Source: EPA Report 12060 EGU, March 1971

Table 3 gives the percentage distribution of different types of dairy plants in respect to knowledge of BOD_5 loading in reference to BOD_5 surcharges and treatment by city or plant. Over 75% of all types of plants discharged wastes to municipal sewer systems except for concentrated milk plants which which had the highest percentage (41%) of its own treatment plants.

Generally, milk, ice cream, milk and ice cream combination and milk, ice cream and cottage cheese plants had treatment facilities for less than 5% of the plants. Knowledge of BOD values in plant wastes varied as a function of the type of plant involved. In all instances where there was knowledge of the wasteload, one or more of the following criteria were met:

    (a) Plant operated its own treatment plant.
    (b) Plant was paying a surcharge on the basis of waste composition.
    (c) Plant had been cited by a government agency for violation or had been notified to reduce wasteloads to a specific level by a specific time.

However, the operation of a waste treatment facility by the plant did not ensure that the plant management knew the wasteloads going into the treatment facility. This was true for 16 separate plants or 18% of those operating waste treatment facilities. One company operating several treatment plants indicated that, "No BOD or SS solids measurements are made of raw or treated wastewater. The efficiency of the treatment plants are determined by visual inspection."

In all but one of the plants, those plants paying a surcharge on BOD levels knew the BOD_5 values for their plant wastes. Generally, BOD analyses for surcharges were made on a quarterly 3 to 7 day survey on a corrected volume basis. In a few instances grab samples were used and charges were being made in several cases on the basis of values obtained over 2 years earlier.

THE NATURE OF THE INDUSTRY

The dairy industry is made up of a large number of, for the most part, relatively small plants scattered primarily through the milk producing sections of the nation. These plants range from single product milk processing or cheese plants to rather complex multiproduct facilities in which milk, cottage cheese, sour cream, ice cream, yogurt, etc., may, for instance, be produced, according to Watson (13).

Since many of the plants are small and located adjacent to municipalities, the usual practice has been to connect these plants into municipal sewer systems. In fact, almost 90% of the plants dispose of their wastes in this manner.

Another reason for this method of waste disposal is the fact that for the most part dairy plant wastes are biodegradable and compatible with the wastewater present in a municipal system. Even fats, oils and greases present in dairy plant wastes are edible and biodegradable so municipalities need not feel the same concern for these materials as is the case with the same constituents of petroleum origin. Since the average size of the dairy plant rules against its being able to afford or professionally operate a pretreatment facility, the industry as a general rule would prefer to buy this sewerage service from the city.

It deserves mentioning here that plant people should exhaust the in-plant, short-of-treatment approach as the soundest and simplest method of controlling a waste problem. In addition to coming to grips with the pollution problem, such action will also result in cost reductions through improved production efficiencies, reductions in losses and reductions in water usage.

The industry has been in a dynamic state for the past several decades with

major changes occurring within the industry. During this period of time,
there has been a slight increase in the amount of milk shipped from farms
for processing, a material decrease in the number of plants processing dairy
foods, a very significant increase in the amount of milk processed by each
plant, and a major trend towards automation and mechanization. These
changes have significantly increased the waste load per plant. Under pre-
vailing practices, waste loads per plant can be expected to range from
2,000 to 10,000 pounds of BOD per day.

The trend over the last several years for reduced milk production appears
to be reversing in that the amount of product being processed has actually
increased from 104 billion pounds in 1960 to 108 billion pounds in 1969,
whereas the total production of milk on the farm changed from 123 billion
pounds in 1960 to 116 billion pounds in 1969.

Trends which have significance to the nature of dairy wastes and waste
treatment include (a) the marked decrease in the number of dairy plants
and increased production per plant, (b) changes in the relative production
of various types of dairy foods with different levels of waste loads, (c) automation of plant processes to an increasing degree with increasing plant size
and consolidation, and (d) a shift in the location of new plant facilities.

Changes in Dairy Plant Size and Product Volume per Plant: Over the past
20 years there has been a marked change in the number and size of many
types of dairy food plants. For cheese and butter plants, a marked decline
in plant numbers occurred prior to 1948. For fluid milk plants, there was
a decrease of about 75% in numbers between 1948 and 1958. Ice cream
plants showed a similar decline. In the past decade, the decline in plants
has been exponential, showing a linear semilog relationship as illustrated
in Figure 2.

Based on these past trends, projected changes in plant numbers up to 1983
are indicated also. Assuming that the number of plants will continue to
decrease at a declining rate, the annual decline in plant numbers for various operations can be projected for the next 30 years as follows.

Type of Plant	Decline in Number of Plants Per Year, %
Cheese	3.9
Ice cream	6.0
Evaporated and condensed milk	4.0
Fluid milk	5.56
Cottage cheese	6.6
Butter	6.2
Dried milk	4.80

The increase in daily production for various plants is illustrated in Figure 3 for the past 20 years, and the predicted growth in plant production is projected to 1983. The relative rate of growth of daily production per plant for butter, cheese, dried milk, evaporated milk and milk are essentially similar, whereas the growth in production of the plant for cottage cheese will occur at about double the rate for other types of dairy product operations. The production of cottage cheese whey per plant is predicted to exceed the per plant production of sweet whey by 1974.

FIGURE 2:   PAST AND PROJECTED TRENDS IN THE NUMBER OF VARIOUS TYPES OF DAIRY PLANTS IN THE UNITED STATES

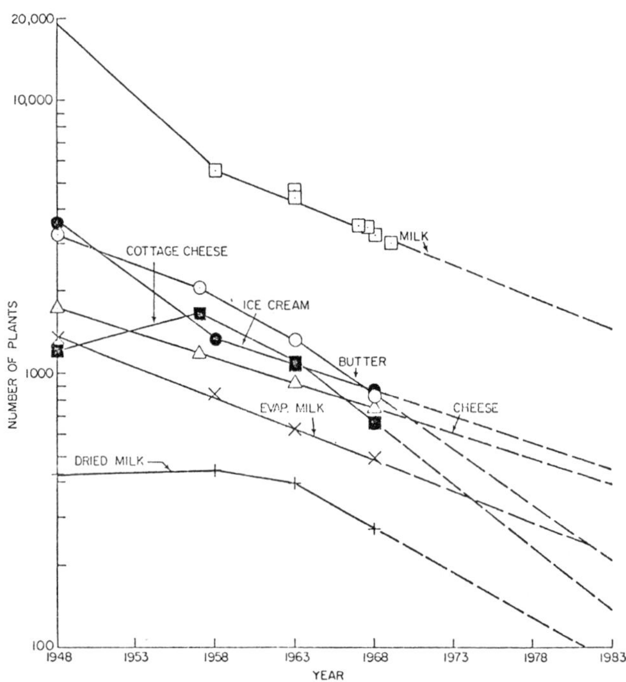

Source: EPA Report 12060 EGU, March 1971

In 1969 the total amount of whey produced was approximately 21 billion pounds with about 6 billion pounds being acid whey. These figures are based on calculated whey volumes from the USDA figures for cheese production. USDA figures for whey production were 17 billion pounds and would appear to be based on incomplete reporting.

FIGURE 3:　PAST AND PROJECTED PRODUCTION PER DAY FOR VARIOUS TYPES OF DAIRY PLANTS IN THE UNITED STATES

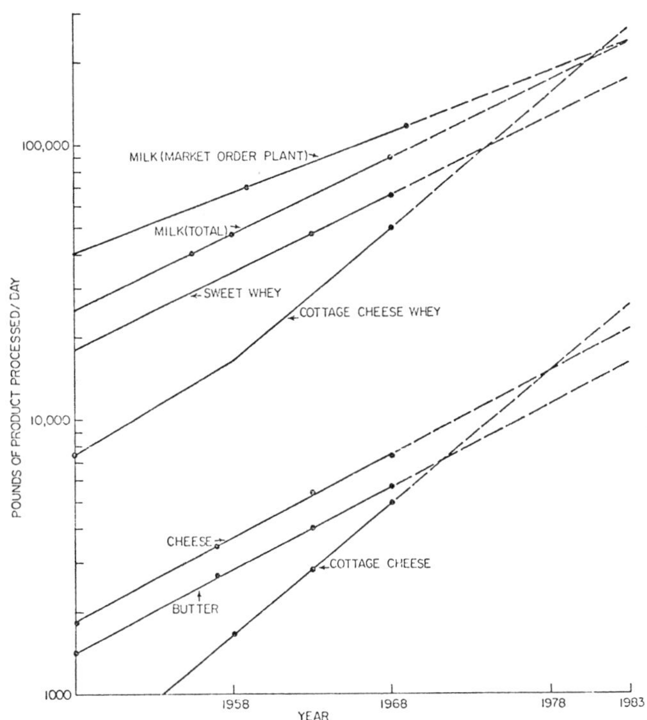

Source:　EPA Report 12060 EGU, March 1971

Whey has long been the most visible pollutant of the dairy food industry. The potential significance of whey and the magnitude of the problem facing the industry in eliminating this material as a waste product is illustrated by the fact that about 20% of the total milk produced in the country is converted into whey at the present time. The very magnitude of the volume of whey contributes materially to the problem.

Cottage cheese whey represents a more serious problem than sweet whey because of its acid nature which limits its utility as a food or feed. Of special concern is the more than 6 billion pounds of cottage cheese washwater approximating 70,000,000 pounds of BOD. The solids are too dilute to recover economically, and the washwater creates problems in waste treatment which will be detailed later in the report.

At the present time, the fluid milk plants, which operate under various
Federal Milk Market Order systems and handle about 50% of the fluid milk
volume, average over 100,000 pounds of milk per day. This could be expected to increase to about 225,000 pounds per day by 1983. Technolog-
ically, maximum plant size appears to be unlimited, but specific limitation
on plant size in the future includes: (a) procurement problems associated
with bringing milk long distances, (b) the problems of distribution over
long distance, (c) manageability of large, complex plants, (d) control of
waste loads, and (e) local sanitary district regulations and taxation policies.

PROCESSES EMPLOYED BY THE INDUSTRY

FLUID MILK PRODUCTION

The definition of Standard Industrial Classification 2026 (SIC-2026),
fluid milk, is as follows: Establishments primarily engaged in processing (pasteurizing, homogenizing, vitaminizing, bottling) and distributing fluid milk and cream and related products, including cottage cheese.

The fluid milk industry has grown steadily over the years and this trend
is expected to continue.

Geographically, population patterns govern plant locations. There is a
trend toward exceptionally large regional plants with distributions over
wide areas. The manufacturing process for fluid milk may be outlined as
shown in the block flow diagram in Figure 4 on the following pages.

1. Receipt — Raw milk is received in tank truck quantities, although a few
smaller plants continue to receive milk in 10-gallon cans.

2. Raw Milk Storage — Raw milk is pumped from receiving to refrigerated
storage tanks until needed. Milk from raw milk storage proceeds to No. 11,
the by-products department, or to clarification.

3. Clarification — Raw milk is clarified (strained) in a centrifugal device,
although in smaller plants mechanical filters may be used.

4. Pasteurization — The clarified milk is usually pasteurized in a continuous flow pasteurizer, although batch type units may be used in smaller
plants.

FIGURE 4:   BLOCK FLOW DIAGRAM FOR FLUID MILK PRODUCTION

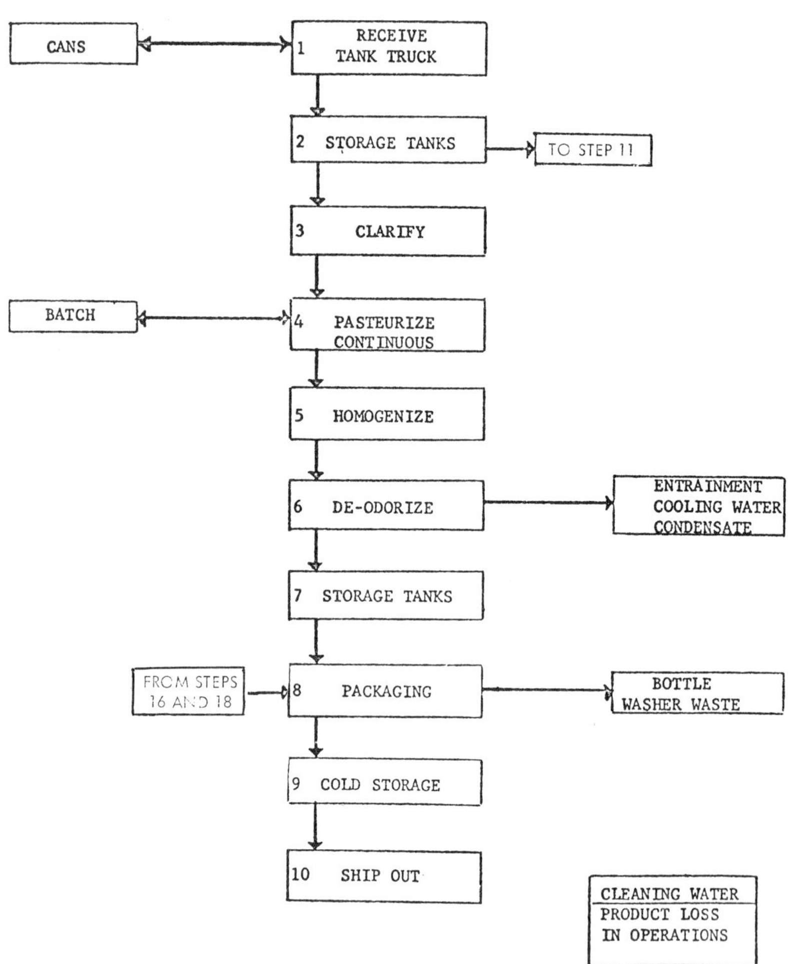

(continued)

FIGURE 4:   (continued)

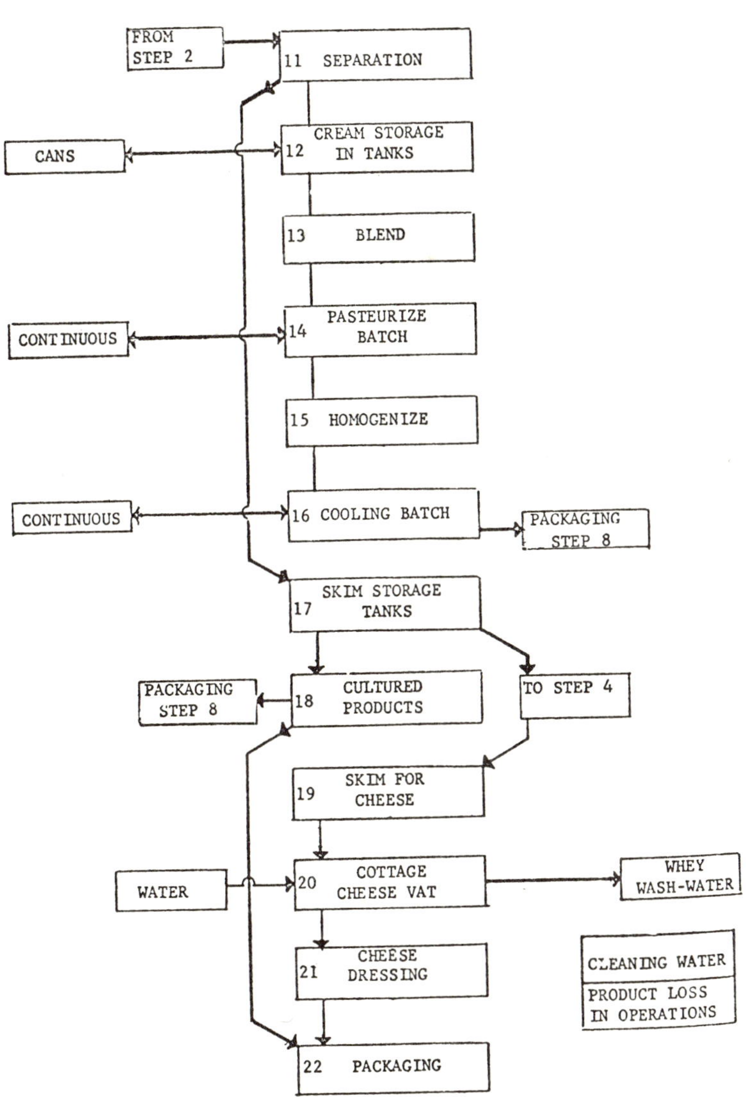

Source: FWPCA Publication IWP-9

5. Homogenization — The pasteurized product is homogenized in a pressure pump which breaks up the butterfat particles to keep them in suspension.

6. Deodorization — The homogenized milk is subjected to a vacuum steam injection treatment to remove off-odors and flavors. Where flavors and odors are not serious, steam may be eliminated and only vacuum treatment used. In some areas the milk supply is of such quality that "grading" can be done during receiving so that deodorization is not necessary.

7. Pasteurized Storage — The pasteurized product is cooled after leaving the previous treatments and is sent to storage tanks and held until needed in packaging.

8. Packaging — Milk is packaged or bottled on automatic machines in a number of different type containers, including glass bottles, paper cartons, plastic bottles, and plastic bags in cardboard box units. In the packaging operation, the packages are usually placed in wire or plastic cases and conveyed to a refrigerated cold storage area.

9. Cold Storage — The packaged product is held in cold storage until needed for shipping.

10. Ship Out — The packaged product is drawn from cold storage and placed into refrigerated trucks for delivery.

Returning to the raw milk storage, one can proceed with the fundamental process for by-products.

11. Separation — The raw milk is separated into cream and skim milk in a centrifugal device and the two products are sent to refrigerated storage.

12. Cream Storage — Cream is normally stored in refrigerated tanks, but occasionally may be stored in cans.

13. Blending — Cream is blended with whole milk and miscellaneous additives placed in the cream at this time to make the various grades of cream for bottling.

14. Pasteurization — The blended products are generally pasteurized in batch quantities, although in larger plants a continuous pasteurizer may be used, such as in No. 4.

15. Homogenization — The pasteurized product is homogenized as in No. 5.

16. Cooling — The homogenized product is cooled in batches, although
in larger plants this may be a continuous process.

17. Skim Milk — Skim milk from the separation process, No. 11, is stored
in refrigerated tanks and is used as follows:

> A portion is returned to the raw milk storage tanks to standard-
> ize the product to a controlled percentage of butterfat.

> A portion is sent to pasteurization, No. 4, and continues
> through Steps 5, 6, and 7, to packaging, No. 8.

> Skim milk is used also in other products, as follows:

18. Cultured Products — Skim milk is processed into buttermilk and yogurt
in batch type processors and sent to packaging, No. 8.

19. Skim Milk for Cheese — Skim milk is drawn from storage tank (17) and
pasteurized as in No. 4, and sent to cottage cheese vats.

20. Cottage Cheese Vat — The pasteurized product is cooled in pasteurizers
to the desired "setting" temperature and pumped into cheese vats. In the
cheese vats the skim milk is innoculated with a bacterial "culture". At the
end of a controlled period of time, the curd resulting from the setting is cut
into small pieces and cooked in the vat.

At the end of the cooking period, the whey is drained off and becomes
available for by-product manufacture or is sent to waste. The curd from
the set-cooking vat is washed with potable water to complete the removal
of whey and to perform a cooling function. This water goes to waste. In
large plants, the washing and draining may occur in a separate piece of
machinery.

21. Cheese Dressing — Cream dressing is made under by-products Steps 13
through 15, and pumped to the cheese and blended. In small plants the
curd and dressings are mixed in cans and stored until packaged.

22. Packaging — The completed cheese is pumped to packaging where it
is placed in containers and sent to (9) cold storage.

The fundamental fluid milk process changed little from 1950 to 1970 and
little change is forecast to 1977. Nevertheless, several developments of
interest have occurred. The most significant change has been in the number
reduction of plants. Due to economical pressures, many small plants have
closed or have merged. This trend is expected to continue.

Since 1950, bulk tank trucks have largely replaced the 10-gallon cans used in Step 1, "Receipt", of the fundamental process. The trend has occurred because the use of trucks has virtually eliminated physical labor, improved sanitation maintenance and reduced the likelihood of contamination.

Self-cleaning (CIP) separators used in Step 3 of the fundamental process are now available. Such machinery reduces the amount of manual washing required, as well as the reduction of physical labor.

Because of tremendous volume, large plants utilize continuous flow equipment, as opposed to batch type machinery. This development has tended to reduce the percentage of plant loss in operations and, consequently, has helped to minimize wastes. Greatly improved heating and refrigeration systems have reduced water needs considerably.

In the early 1950's, vacuum deodorizing equipment became available and is now used in many areas to eliminate feed, onion and other off-flavors. This equipment has tended to increase plant product losses.

The trend in packaging is to smaller units which better serve the needs and desires of the consumer. Automatic packaging continues to replace manual methods. Not only is the amount of waste reduced, but new machinery fills more accurately.

The processing of certain cultured products is the only significant process change to occur since 1950. In the processing of sour cream, the use of chemical means as opposed to biological cultures is used in a small way and reduces the time of the process, but does not change the amount of waste. Research work is underway to develop continuous setting methods. However, commercial production does not appear imminent.

Hot pack sour cream and cottage cheese is now in use in a small way. In this process the packages are filled with the product prior to culture growth. The growth takes place in the package, and thus eliminates this step in the batch process, thus theoretically reducing waste. The actual plants utilizing this method are experiencing increased waste during the technological development of machinery.

Permanent stainless steel piping systems were introduced in the early 1950's. Such systems are cleaned in place, as opposed to the daily take-apart systems formerly accepted. This type equipment reduces the quantity of soap required and, therefore, reduces waste. The fact that the systems are permanently installed has reduced plant product losses; also, sanitation and product shelf life have been increased, a factor which has tended to reduce waste.

Significant changes have occurred in material handling within plants by the introduction of sophisticated conveyors and stacking, grouping and palletization equipment. Even though machines have tended to increase individual plant wastes through the enlarged usage of water-soap lubricants, product loss and waste has been reduced because of less likelihood of package damage.

A large amount of dairy product manufacture has been replaced by non-dairy products especially in coffee creamers in which vegetable fat is substituted for butterfat. The trends may best be shown in tabular form, which follows. The reader should note that the alternative subprocesses and other industry changes have occurred over a span of years. The process which will become prevalent is identified as (P), and that which is becoming less used as (S). The estimates represent the observations and opinions of people in the industry.

TABLE 4:   ESTIMATED PERCENTAGE OF PLANTS EMPLOYING PROCESS

	1950	1963	1967	1972	1977
P Receive in tank trucks	0	50	60	70	90
S Receive in cans	100	50	40	30	10
P Centrifuge manually	100	100	99	92	85
S Clean-in-place	0	0	1	8	15
P Pasteurize continuously	30	50	60	70	90
S Pasteurize batch	70	50	40	30	10
P Deodorizer installed	0	10	15	20	25
S Not installed	100	90	85	80	75
P Package automatically	70	80	85	90	92
S Package manually	30	20	15	10	8
P Sour cream - biologically	100	100	98	96	90
S Sour cream - chemically	0	0	2	4	10
P Batch set cottage cheese	100	100	100	98	95
S Continuous set cottage cheese	0	0	0	2	5
P Cold pack cultured products	100	100	99	97	93
S Hot pack cultured products	0	0	1	3	7
P Welded piping	0	20	40	60	95
S Take-apart piping	100	80	60	40	5
* P Automatic material handling	0	60	50	65	85
S Manual material handling	100	40	50	35	15

*Almost all plants have conveyors of some type. This heading indicated utilization of casers, stackers, palletization devices, etc.

Source: FWPCA Publication IWP-9

The subprocesses shown in Table 4 do not require different type treatment from the fundamental processes; however, the choice of subprocesses is largely determined by the total volume produced. The continuous flow processes tend to have less waste per pound of finished product because of the greater productivity per pieces of equipment.

The whey from cottage cheese manufacture, product spillage and waste during normal processing, and cleaning water and soaps represent the significant wastes for all processes and subprocesses. In order to best estimate total industry waste and wastewater, it is desirable to identify levels of technology within the industry. The table below illustrates three technological levels. The fundamental process steps from Figure 4 are used as reference.

TABLE 5:   COMPARATIVE TECHNOLOGY

	Older Technology	Typical Technology	Advanced Technology
1.	Receive products in 10-gallon cans	Receive almost all products in tank trucks, although a certain amount in 10-gallon cans	Receive all products in tank truck quantities
2.	Store in refrigerated tanks	Store in refrigerated tanks	Store in refrigerated tanks
3.	Clarify, using strainers	Clarify in a centrifugal device	Clarify in a centrifugal device
4.	Pasteurize in batch quantities	Pasteurize in a continuous manner	Pasteurize in a continuous manner
5.	Homogenize in a continuous pressure pump	Homogenize in a continuous pressure pump	Homogenization in a continuous pressure
6.	Deodorization not used	Deodorize in steam vacuum equipment	Deodorize in steam vacuum equipment
7.	Pasteurized storage in surge tanks, and (8)	Pasteurized refrigerated storage tanks	Pasteurized refrigerated storage tanks
8.	Packaging in small commercial size automatic or semiautomatic machinery	Package in automatic machinery, place containers into cases manually and stack manually on slow speed lines and automatically on high speed lines	All packaging takes place on fully automatic machinery and finished packages placed in cases automatically and stacked automatically
9.	Product sent to cold storage and handled in case quantities	Cold storage, where product is inventoried in stack quantities	Product sent to cold storage where stacked cases are handled in unit or pallet quantities
10.	Ship out in cases handled on an individual basis or stacks	Shipped out in stack quantities	Shipped out in unit or pallet quantities
	BY-PRODUCTS:		
11.	Separation for by-products occurs in a centrifugal device	Separation for by-products performed in a centrifugal device	Separation in the advanced plant will take place in a separator located within the pasteurizer (4), so arranged that the milk blends, and skim milks come out of the pasteurizer in two streams already completely pasteurized and homogenized at the correct fat content to be sent directly to (7), pasteurized storage tanks, and then to packaging. The skim milk out of the advanced pasteurizer would be sent to (18)
12.	Cream stored in 10-gallon cans	Cream sent to refrigerated storage tanks	
13.	Blending of various creams takes place in small vats, usually the same vat as (14)	Blending takes place in refrigerated storage tanks	
14.	Batch pasteurization	Pasteurization of by-products would occur in a smaller plant in batch quantities and in a large plant on a continuous pasteurizer	
15.	Homogenization on a continuous pressure pump	Homogenization in a continuous pressure pump	
16.	Cooling in batch quantities	Cooling in a continuous manner and then product pumped to pasteurized surges (7) for packaging	
17.	Skim milk for the separation process may be stored in 10-gallon cans in a small plant and in small vats in larger plants	Skim milk from the separator is normally sent directly to storage tanks for later use	

(continued)

TABLE 5:   (continued)

Older Technology	Typical Technology	Advanced Technology
CULTURED PRODUCTS:		
18. Cultured products in small plants made in 10-gallon cans and the larger plant in batch quantities	For cultured products skim milk is pumped directly to the batch-type processors	Skim milk out of advanced pasteurizer is sent to the cultured products batch processors, or (19)
19. Skim for cheese would be pumped directly from the batch pasteurizer to (20), the cottage cheese vat	Skim for cottage cheese would be pumped directly to the cottage cheese vat from continuous pasteurizer	The cottage cheese vats at the correct temperature for innoculation
20. Setting, cooking, draining whey and washing of curds would all take place in cottage cheese vat, which would be a manual type	Cottage cheese vat is equipped with mechanical agitation and pushers; however, curd cutting, cooking and whey draining and washing would normally occur in this vat, as will (21)	Product will be set, cut and cooked in cottage cheese vat; however, the mixture of curd whey will be pumped to a separate draining device which will also have provision for washing and will be located on a weighing device so that cheese dressing may be applied
21. Cheese dressing would be made in a batch processor and mixed with the cheese in 100# cans, stored until cream is absorbed by the curd and is ready for packaging	Application of cheese dressing	Cheese dressing may be applied in known weight quantities and mixed mechanically at this point. The finished product will be pumped to (22)
22. Product transferred by hand to semi-automatic machinery or filled by hand and finished product sent to (9), cold storage	Products from the cheese vats will be pumped to semiautomatic packaging machinery where it is manually cased and sent to (9), cold storage	Automatic packaging machinery from which it will be packed automatically and sent in conveyor quantities or pallet quantities to cold storage (19)
23. Take-apart piping	Partial CIP piping	CIP piping
24. Manual material handling	Partial automatic material handling	Automatic material handling

Source: FWPCA Publication IWP-9

CREAMERY BUTTER PRODUCTION

The definition of Standard Industrial Classification 2021 (SIC-2021), creamery butter, is as follows: Establishments primarily engaged in manufacturing creamery butter.

Butter production has for many years been declining, as a result of competition from oleomargarine.

Per capita consumption reached a low in 1966, but total porduction is now expected to increase in proportion to population growth. The manufacturing process of butter may be outlined as shown in the block flow diagram in Figure 5 on the following page.

1. Receipt — Raw (unpasteurized) milk and cream are received from the farm in either tank trucks or 10-gallon cans.

2. Storage — The contents are subsequently pumped to refrigerated storage tanks. (A plant may have a refrigerated storage room where the milk and

cream remain before being dumped into storage tanks.)

3. Separation — From the storage receptacles, the raw products are passed through a heater. The raw milk is warmed to a temperature of 90°F. and then centrifuged. The cream with a butterfat content of 30 to 40% is separated and stored separately. (The remaining skim milk is available for by-product use.)

FIGURE 5:   BLOCK FLOW DIAGRAM FOR CREAMERY BUTTER PRODUCTION

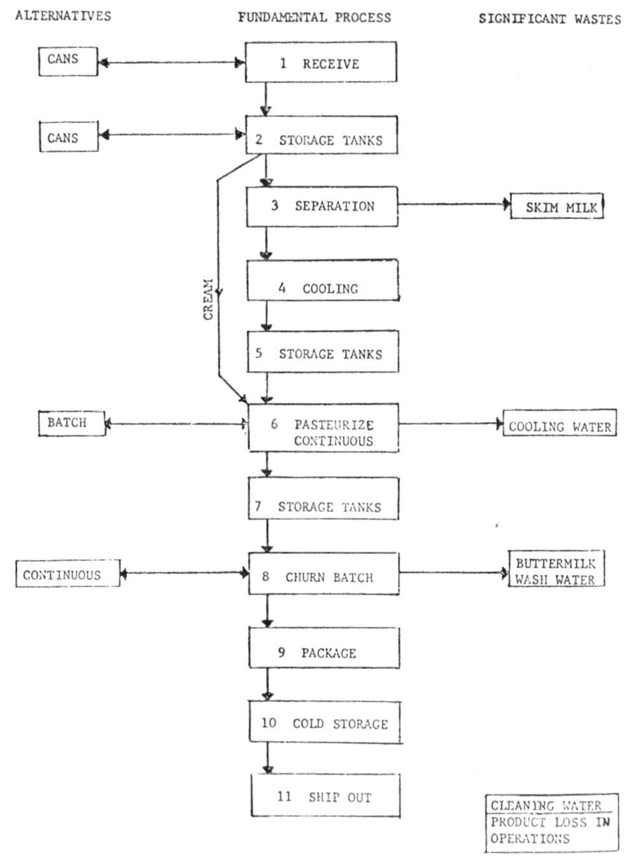

Source: FWPCA Publication IWP-9

4. Cooling — The cream is cooled in continuous coolers and pumped to storage.

5. Storage — The resultant product is held in tanks under controlled refrigeration.

6. Pasteurization — The raw cream is next pumped to a continuous flow
pasteurizer, where the liquid is pasteurized and cooled. Small plant concerns may continue to use the vat type pasteurizer.

7. Pasteurized Storage — The cooled product from the pasteurizers is stored
in tanks awaiting utilization further in the process stream.

8. Churning — The pasteurized cream is tempered to 45°F. and is churned.
The buttermilk resulting from the churning process is drained; the butter
granules are washed, drained, rewashed, drained, standardized to 80% fat
with addition of water, salt, color and flavoring, and "worked" to the desired consistency. The butter portion is sent to packaging. The buttermilk portion becomes available for by-product use or is wasted.

9. Packaging — Butter is placed in various types of packaging machinery,
where the commodity is extruded to the desired shape, wrapped and packaged.

10. Cold Storage — Butter is placed in cold storage until needed for customer delivery.

11. Shipping — Packaged butter is usually placed in refrigerated vehicles
for delivery to customers.

The fundamental butter process has changed little over the years, and little
change is forecast to 1977. Nevertheless, several developments of interest
have occurred.

The trends may best be shown in tabular form as shown on the following
page. The reader should note that these industry changes have occurred
over a span of years.

The process which will become prevalent is identified as (P), and that
which is becoming less used as (S). The estimates represent the observations
and opinions of people in the industry including processors, material and
equipment suppliers and manufacturers and industry associations and consultants.

The subprocesses shown in Table 6 do not require different treatment from
the fundamental processes; however, the choice of subprocess is largely
determined by the total volume produced. Large plants often utilize continuous flow processes because of greater productivity per piece of equipment. These processes generate less waste per pound of finished product.

Skim milk, buttermilk, product spillage, cleaning water and soaps, all
constitute the significant wastes for any type process utilized.

TABLE 6:   ESTIMATED PERCENTAGE OF PLANTS EMPLOYING PROCESS

		1950	1963	1967	1972	1977
P	Receive in tank trucks	0	40	50	60	70
S	Receive in cans	100	60	50	40	30
P	Separate manually	100	100	98	96	93
S	Separate CIP	0	0	2	4	7
P	Pasteurize continuously	0	20	25	30	35
S	Pasteurize batch	100	80	75	70	65
P	Churn batch	100	100	100	98	95
S	Churn continuously	*	*	*	2	5
P	Package automatically	15	40	50	60	70
S	Package manually	85	60	50	40	30
P	CIP piping	0	20	30	40	60
S	Take-apart piping	100	80	70	60	40
P	Automatic material handling	20	50	60	70	75
S	Manual material handling	80	50	40	30	25

*Less than 1%

Source: FWPCA Publication IWP-9

To best estimate total industrial waste and wastewater, the existing levels
of technology are identified.   The table below shows three technological
levels.   The fundamental process steps used are taken from Figure 5.

TABLE 7:   COMPARATIVE TECHNOLOGY

	Older Technology	Typical Technology	Advanced Technology
1.	Receive in cans	Receive in tank trucks	Receive in tank trucks
2.	Store in cans	Store in tanks	Store in tanks
3.	Heat, then separate centrifugally	Heat, then separate centrifugally	Heat, then separate centrifugally
4.	Cool in batches	Cool continuously	Cool continuously
5.	Store raw product in cans	Cold storage in tanks	Store raw products in tanks
6.	Pasteurize and cool in batches	Pasteurize and cool in batches	Pasteurize and cool in batches
7.	Storage in batch pasteurizers	Pasteurized storage in tanks	Pasteurized storage in tanks
8.	Churn in batches	Churn in batches	Churn continuously
9.	Package manually	Package semi-automatically	Package automatically
10.	Store in cold storage	Inventory in cold storage	Inventory in cold storage
11.	Ship out	Ship out	Ship out
12.	Take-apart piping	Partial CIP piping	CIP piping
13.	Manual material handling	Partial automatic material handling	Automatic material handling

Source: FWPCA Publication IWP-9

CHEESE MANUFACTURE

The definition of Standard Industrial Classification 2022 (SIC-2022), cheese, natural and processed, is as follows: Establishments primarily engaged in manufacturing all types of natural cheese (except cottage cheese, Industry 2026), processed cheese, cheese food and cheese spreads.

The cheese industry has grown over the years at a slightly faster rate than population growth. This trend is expected to continue. The manufacturing process of cheese is shown in the block flow diagram in Figure 6.

FIGURE 6: BLOCK FLOW DIAGRAM FOR NATURAL AND PROCESSED CHEESE MANUFACTURE

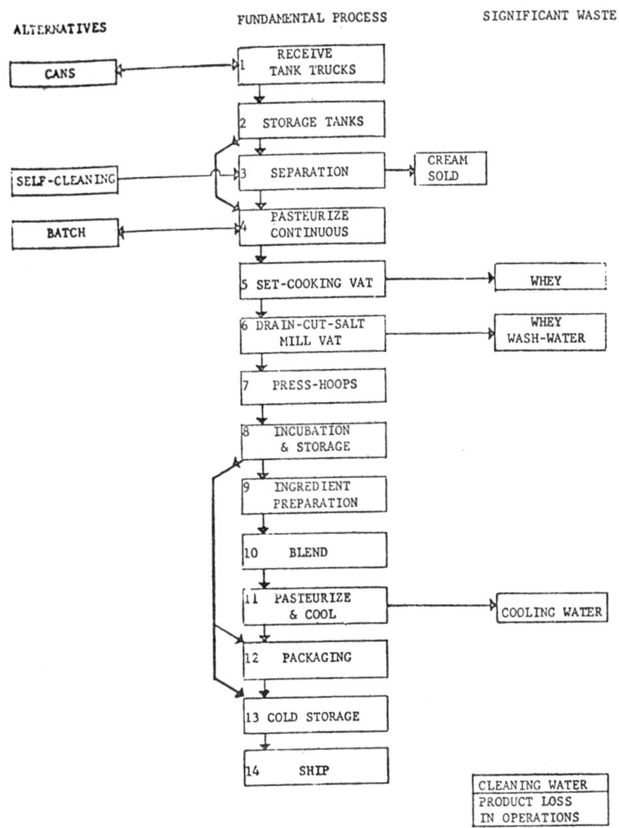

Source: FWPCA Publication IWP-9

1. Receipt — Raw milk and skim milk are received in tank trucks and are emptied by pumping to storage.

2. Storage — The raw (unpasteurized) milk is stored in refrigerated tanks until ready for further use.

3. Separation — For low fat cheese, the raw products are pumped through a heating device and sent to a centrifugal separator which removes all or part of the cream from the product. This cream becomes available for by-product manufacture.

4. Pateurization — The raw milk is usually pasteurized in a continuous flow pasteurizer, although in smaller operations batch pasteurizers continue to be used.

5. Batch Set-Cooking — The pasteurized product is normally cooled in the pasteurizer to the desired temperature and pumped into cheese vats. The milk is then innoculated with a culture. At the end of a controlled period of time, the curd which results from bacterial action is drained and becomes available for either by-product manufacture or is treated as waste.

6. Batch Drain-Cut-Salt-Mill Vat — The curd from the cooking vat is washed with potable water which completes the rinsing away of whey and serves as a cooling medium. This water goes to waste. At this time salt may be added, and the curd may be cut or milled.

7. Press-Hoops — The curd is pressed (compressed) and placed into hoops, which are can-shaped molds.

8. Incubation and Storage — The cheese hoops are placed in controlled environment storage rooms to permit "aging", the incubation period neces-sary to complete the formation of cheese. This period may be a very short time or it may be a matter of months, the time depending upon the variety of cheese being manufactured. When the cheese is removed from the in-cubation storage a portion may go directly to packaging while the other portion may be used for processed cheese.

9. Ingredient Preparation — In the preparation of processed cheese, the hoop cheese is ground and placed in vats where stabilizers, flavoring, and other needed ingredients are added.

10. Blending — The ingredients are then blended.

11. Vat Pasteurization and Cooling — The blended ingredients are pasteur-ized, partially cooled, and sent to packaging.

12. Packaging — The hoop and processed cheese are conveyed to filling
and packaging machines which shape and place the cheese in characteristic packages and wrappers.

13. Cold Storage — From packaging, the cheese is stored and inventoried
in a cold storage area until needed.

14. Shipping — The cheese is removed from cold storage and placed on refrigerated vehicles for delivery to the consumer.

The fundamental cheese process has changed little in recent years from
1950 to 1966, and little change is forecast to 1977. Nevertheless, several developments of interest have occurred. The trends may best be shown
in tabular form, which follows. The reader should note that the alternative
subprocesses and other industry changes have occurred over a span of years.

The process which will become prevalent is identified as (P), and that which
is becoming less used as (S). The estimates represent the observations and
opinions of people in the industry, including processors, material and equipment suppliers and manufacturers and industry associations and consultants.

TABLE 8:　ESTIMATED PERCENTAGE OF PLANTS EMPLOYING PROCESS

		1950	1963	1967	1972	1977
P	Receive in Tank Trucks	-0-	40	50	60	70
S	Receive in Cans	100	60	50	40	30
P	Separator (Manual)	100	100	98	96	94
S	Separator CIP	-0-	-0-	2	4	6
P	Pasteurize Continuously	-0-	-0-	2	4	7
S	Pasteurize Batch	100	100	98	96	93
P	Batch Set	100	100	99	98	95
S	Continuous Set	-0-	-0-	1	2	5
P	Package Automatically	10	35	50	65	75
S	Package Manually	90	65	50	35	25
P	CIP Piping	-0-	20	30	40	60
S	Take-apart Piping	100	80	70	60	40
P	Automatic Material Handling	20	50	60	70	75
S	Manual Material Handling	80	50	40	30	25

Source:　FWPCA Publication IWP-9

The subprocesses shown above in Table 8 do not require different treatment
from the fundamental processes; however, the choice of subprocess is largely
determined by the total volume produced. Large plants often utilize continuous flow processes because of greater productivity per piece of equip-

ment. These processes generate less waste per pound of finished product. The whey from the cheese manufacturing process, wash water, product spillage and waste during normal processing, and cleaning water and soaps represent the significant wastes for all processes and subprocesses. Whey constitutes by far the largest volume of waste and is high in protein content as well as acidity.

In order to best estimate total industry waste and wastewater, it is desirable to identify levels of technology within the industry. The table below illustrates three technological levels. The fundamental process steps from Figure 6 are used as reference for this table.

TABLE 9:   COMPARATIVE TECHNOLOGY

Older Technology	Typical Technology	Advanced Technology
1. Receive in cans	Receive in tank trucks	Receive in tank trucks
2. Store in cans	Store in tanks	Store in tanks
3. Separation, if required, centrifugally	Separate as required; heat and separate centrifugally	Separate as required; heat and separate centrifugally
4. Pasteurize in batches	Pasteurize continuously	Pasteurize continuously
5. Set-cooking vat manually agitated	Set-cooking vat equipped with mechanical agitation and pushers	Set-cooking vat equipped with mechanical agitators and pushers
6. Drain, cut, salt and mill in the set-cooking vat	Drain, cut, salt, mill vat, curd pumped to this vat from the set-cooking vat	Drain, cut, salt, mill vat, curd pumped to this vat from the set-cooking vat
7. Press in hoops manually	Press in hoops, curd conveyed and pressed automatically	Press in hoops, curd conveyed and pressed automatically
8. Incubate in controlled environment	Incubation in storage under controlled enviroment	Incubation in storage under controlled enviroment
9. Ingredient preparation for processed cheese, manual	Ingredient preparation equipment usually mechanical	Ingredient preparation equipment usually mechanical
10. Blend ingedients manually	Blend ingredients mechanically	Blend ingredients mechanically
11. Pasteurize and cool in batches	Vats, batch pasteurization and cooling	Vats, batch pasteurization and cooling
12. Package manually	Package in large part automatically	Package automatically
13. Inventory in cold storage	Inventory in cold storage	Inventory in cold storage
14. Ship out	Ship out	Ship out
15. Take-apart piping	CIP piping (partial)	CIP piping
16. Manual materials handling	Partial automatic materials handling	Automatic materials handling

Source: FWPCA Publication IWP-9

ICE CREAM MANUFACTURE

The definition of Standard Industrial Classification 2024 (SIC-2024), ice cream and frozen desserts, is as follows: Establishments primarily engaged in manufacturing ice cream and other frozen desserts.

The ice cream industry has grown steadily over the years and this pattern is expected to continue. Geographically, the plant locations reflect population patterns.

There is a trend towards exceptionally large regional plants with distributions over wide areas. The manufacturing process for ice cream may be outlined as shown in the block flow diagram in Figure 7.

FIGURE 7:　BLOCK FLOW DIAGRAM FOR ICE CREAM AND FROZEN DESSERT MANUFACTURE

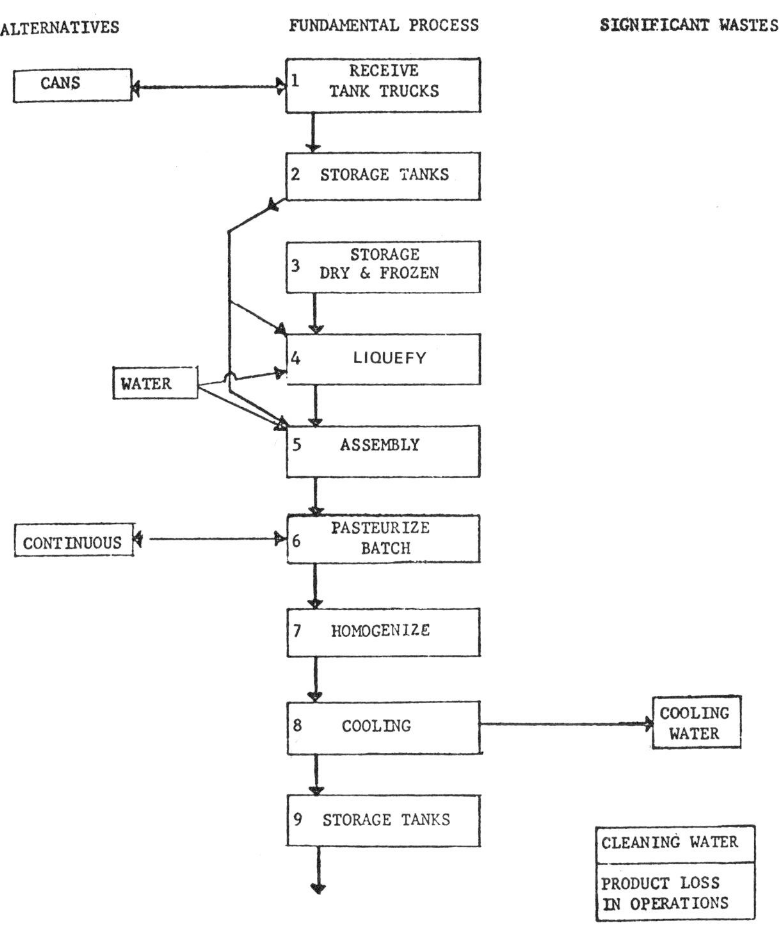

(continued)

FIGURE 7:   (continued)

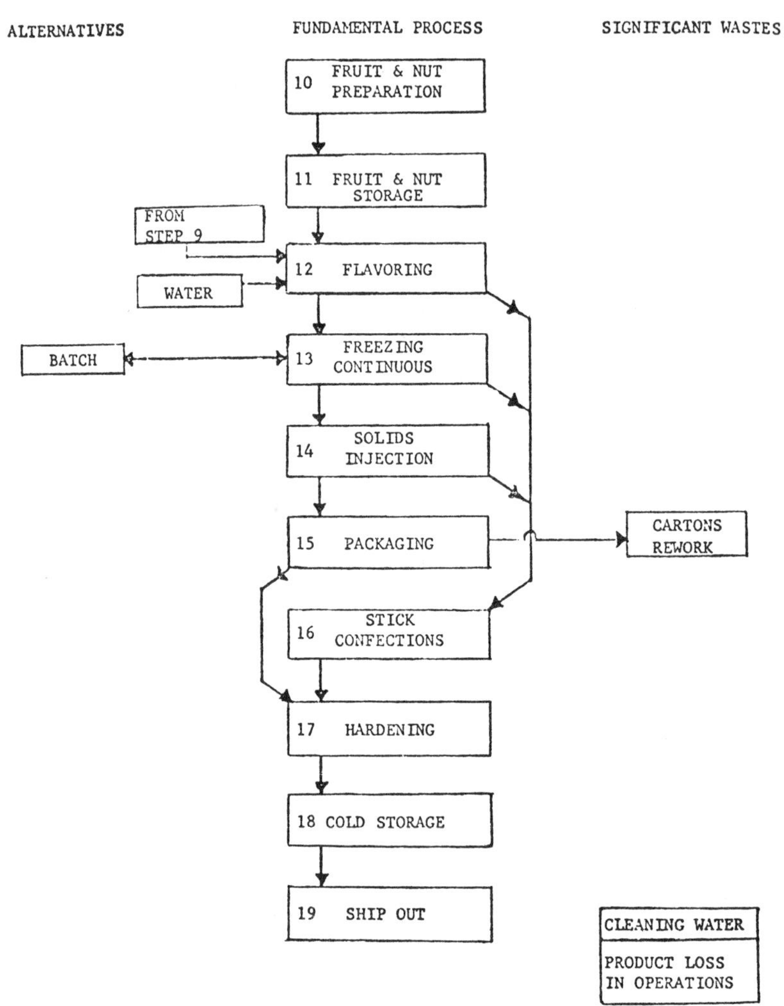

Source: FWPCA Publication IWP-9

1. Receipt — The largest volumes of products received are liquid cream, liquid condensed milk, whole milk, corn syrup and cane sugar syrup. These are normally received in tank truck quantities in liquid form, although in smaller operations cream and condensed milk may be received in cans, and corn and cane sugars in dry form.

2. Raw (Unpasteurized) Product Storage — The dairy products are normally

stored in refrigerated tanks and the sugar syrups in heated tanks.

3. Dry and Frozen Storage — Ingredients such as stabilizers, emulsifiers, and chocolate powders, as well as miscellaneous quantities of sugars are stored in dry form usually in drums and bags. Cream and butter are often purchased during the surplus season of the year in the frozen form and stored frozen. Similarly, various fruits are stored frozen in subzero rooms.

4. Liquefy — The dry and frozen ingredients are normally converted to liquid solution prior to further use. This is done in a high speed blending device in which the dry and frozen ingredients are mixed thoroughly with water or milk.

5. Assembly — The liquified ingredients, as well as the liquid dairy products and sugar syrups are assembled in batch quantities according to formulae.

6. Mix Pasteurization — Upon assembly, a given batch of mix is pasteurized. The normal method is in batch quantities, although in larger plants a continuous pasteurizer may be used.

7. Homogenization — After pasteurization is completed, the mix is homogenized in a high pressure pump which breaks up fat particles so that they will stay in suspension in the finished product.

8. Cooling — The warm mix from the homogenizer is cooled to 40°F. or lower in a continuous cooler.

9. Pasteurized Storage — The cold pasteurized mix is held in refrigerated storage tanks until needed in the flavoring and freezing operations, No. 12 and No. 13.

10. Fruit and Nut Preparation — Frozen nuts are drawn from storage and roasted, and frozen fruits are drawn from storage, defrosted, and separated into pulp and juice.

11. Fruit and Nut Storage — The pulp fruit and the separate juice is stored in containers under refrigeration for later use in No. 12.

12. Flavoring — Mix is drawn from storage tanks, No. 9, into small mixing vats in which the liquid fruit juices are added for flavoring. At this time artificial flavors may also be added.

13. Freezing — The flavored mixes are pumped to ice cream freezers which are the industrial type of the familiar frozen custard stand freezers. In the freezer the mix is frozen on the surface of a refrigerated tube and is scraped

off with sharp blades rotating at high speeds which also whip air into the mix to give it its characteristics as ice cream. Freezers are a continuous type device other than in very small plants in which batch type freezers may be used.

14. Solids Injection — The partially frozen ice cream from the freezers passes through a machine which injects nuts and fruit pulp into the stream. This partially frozen ice cream is sent to packaging equipment.

15. Packaging — The packaging machinery which is normally automatic, forms the container, injects a controlled amount of the product, and seals the package.

16. Stick Confections — A special class of packaging-freezing device is the stick confection unit. The product from the ice cream freezer, No. 13, or flavored water base mixes from No. 12, are placed in the stick confection freezer and frozen, sticks inserted, coatings applied, and the finished product packaged in a paper bag or wrapper.

17. Hardening — The partially frozen ice cream is conveyed in packages to a "hardening area" in which the product is subjected to low temperature air circulation and the freezing cycle is completed. The hardening area in a larger plant is usually a continuous "tunnel" device, cycled with the packaging; however, in smaller operations the product may be stacked in the partially frozen form in racks or on shelves and hardened in the storage area.

18. Cold Storage — The hardened ice cream is held in cold storage as inventory until ready for shipment.

19. Ship Out — The hardened ice cream is drawn from inventory and placed on refrigerated route trucks for delivery to the consumer.

The fundamental ice cream and frozen desserts process changed little in recent years, and little change is forecast to 1977. Nevertheless, several developments of interest have occurred.

The trends may best be shown in tabular form as shown on the following page. The reader should note that the alternative subprocesses and other industry changes have occurred over a span of years.

The process which will become prevalent is identified as (P), and that which is becoming less used as (S). The estimates represent the observations and opinions of people in the industry. The subprocesses shown in Table 10 do not require different treatment from the fundamental processes; however, the choice of subprocess is largely determined by the total volume produced. Large plants often utilize continuous flow processes because of greater productivity per piece of equipment.

TABLE 10: ESTIMATED PERCENTAGE OF PLANTS EMPLOYING PROCESS

		1950	1963	1967	1972	1977
P	Receive in Tank Trucks	-0-	25	30	40	60
S	Receive in Cans	100	75	70	60	40
P	Liquefy Manually	100	95	90	85	75
S	Liquefy by Machine	-0-	5	10	15	25
P	Assemble Manually	100	98	97	95	90
S	Assemble Automatically	-0-	2	3	5	10
P	Pasteurize Batch	100	98	97	95	90
S	Pasteurize Continuously	-0-	2	3	5	10
P	Freeze Continuously	75	50	65	80	90
S	Freeze Batch	25	50	35	20	10
P	Automatic Packaging	10	50	65	80	90
S	Manual Packaging	90	50	35	20	10
P	Hardening in Storage	100	99	97	75	50
S	Hardening in Tunnel	-0-	1	3	25	50
P	CIP Piping	-0-	20	30	40	60
S	Take-apart Piping	100	80	70	60	40
P	Auto. Material Handling	-0-	2	3	5	10
S	Manual Material Handling	100	98	97	95	90

Source: FWPCA Publication IWP-9

To best estimate total industrial waste and wastewater, the existing levels of technology are identified. Table 11 below shows three technological levels. The fundamental process steps used are taken from Figure 7.

TABLE 11: COMPARATIVE TECHNOLOGY

Older Technology	Typical Technology	Advanced Technology
1. Receive product in 10-gallon cans	Receive almost all of product in tank truck quantities but continuing to receive a small amount of canned products	Receive all products in tank truck quantities and (2)
2. Store all products in 10-gallon cans	Store all products in refrigerated tanks	Store in tanks
3. Storage of dry and frozen products in bag or unit quantities	Storage of dry and frozen ingredients, most products in bag or unit quantities except the liquid sugars and corn syrups would be utilized	Storage of dry and frozen ingredients at an absolute minimum. Almost all products stored in tank quantities
4. Liquefication of dry and frozen ingredients manually	Liquefication performed in (5)	Liquefication performed in an automatic machine
5. Assemby of ingredients directly into (6)	The assembly vat. Product pumped from assembly vat into (6)	Assembly takes place in a programmed automatic vat on weigh scales and pumped to (6)
6. Batch type mix pasteurizers	Batch pasteurizers	Continuous mix pasteurizers
7. Homogenize in a high speed pressure pump	Homogenization in a high pressure pump continuously	Homogenization in a continuous high pressure pump

(continued)

TABLE 11:   (continued)

Older Technology	Typical Technology	Advanced Technology
8. Cool in batch quantities	Cooling performed in a continuous manner with ammonia DX	Cooling with regeneration in a continuous manner, and product pumped to (9)
9. Store pasteurized products in 10-gallon cans	Pasteurized product stored in refrigerated tanks	Refrigerated storage tanks
10. Fruit and nut preparation manually and product stores in (11)	Fruit and nut preparation performed with machine assistance	Fruit and nut preparation essentially manual with semi-automatic machine assistance
11. 10-gallon cans or unit quantities	Stored in 10-gallon cans or unit quantities	Fruit juices stored in tanks and pumped, and nuts stored in wheeled containers
12. Flavoring performed in 10-gallon cans or in (13)	Flavoring performed in small mixing vats adjacent to (13)	Flavoring performed in either large pasteurized storage vats (9) or in small mixing vats adjacent to (13)
13. Batch freezers	The automatic continuous freezers	The continuous automatic ice cream freezers
14. Solids injected directly into barrel or batch freezers	Solids injected by machine	Solids injected by machine
15. Packaging performed by hand	Packaging in some automatic machinery and some still performed by hand	Packaging in all automatic machinery
16. Stick confections made on manual devices	Stick confections made in small size semiautomatic machinery and hand packaged	Stick confections made and packaged on all automatic machinery
17. Hardening performed by stacking product on shelves or in wire baskets in the open storage room in the same place that (18) is held	Hardening takes place in the same area as inventory storage (18)	Hardening performed in an automatic "freezing tunnel" and product conveyed to (18)
18. Inventory is held	Inventory performed by placing product in wire baskets or on shelves in this area	Cold storage where it is handled in a palletized manner
19. Ship out	Ship out	Product shipped out in pallet loads or some other form of unit load quantities
20. Take-apart piping	Partial CIP piping	CIP piping
21. Manual materials handling	Partial automatic materials handling	Automatic materials handling

Source: FWPCA Publication IWP-9

CONDENSED AND EVAPORATED MILK MANUFACTURE

The definition of Standard Industrial Classification 2023 (SIC-2023), condensed and evaporated milk, is as follows: Establishments primarily engaged in manufacturing condensed and evaporated milk and related products, including ice cream and ice milk mix, and dry milk products. Condensed and evaporated milk production has been declining since 1955 with a marked drop in 1965 and 1966. The trend is expected to continue.

Total farm milk production has been declining, while fluid milk, cheese and ice cream production has increased. With the exception of that portion of condensery production used for ice cream mix manufacture, the

condensery industry operates by utilizing the milk remaining after the requirements of the other segments of the industry have been met. The manufacturing process for condensed and evaporated milk may be outlined as
shown in the block flow diagram in Figure 8.

FIGURE 8:   BLOCK FLOW DIAGRAM FOR CONDENSED AND
EVAPORATED MILK MANUFACTURE

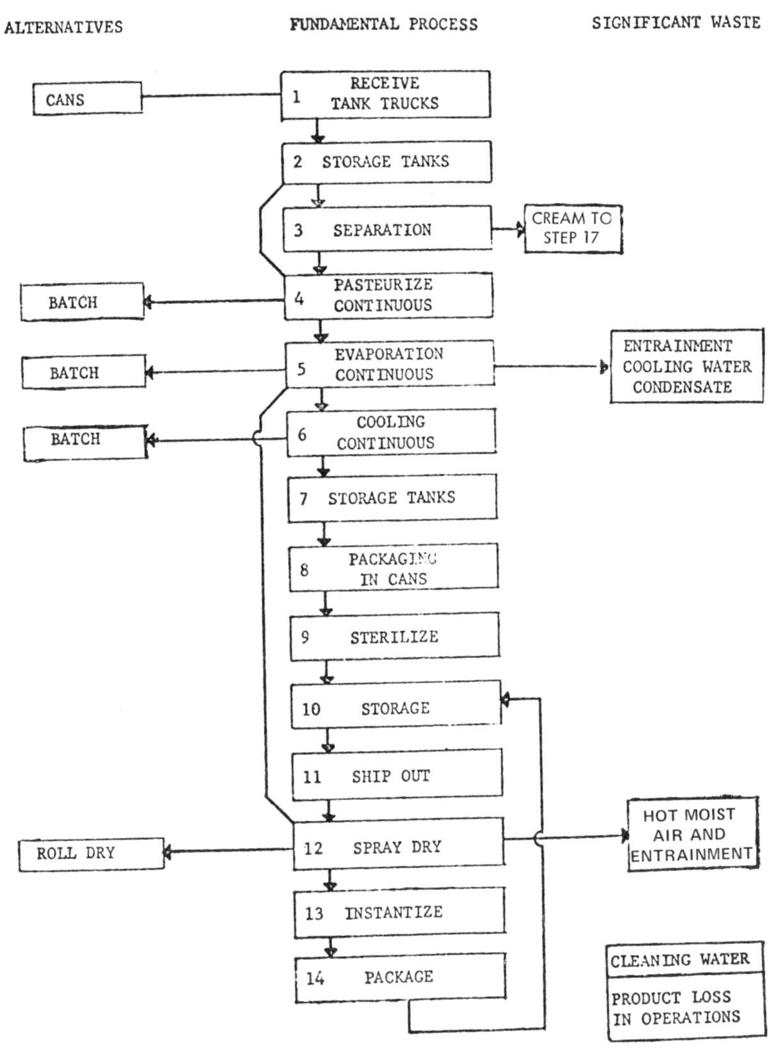

(continued)

FIGURE 8:   (continued)

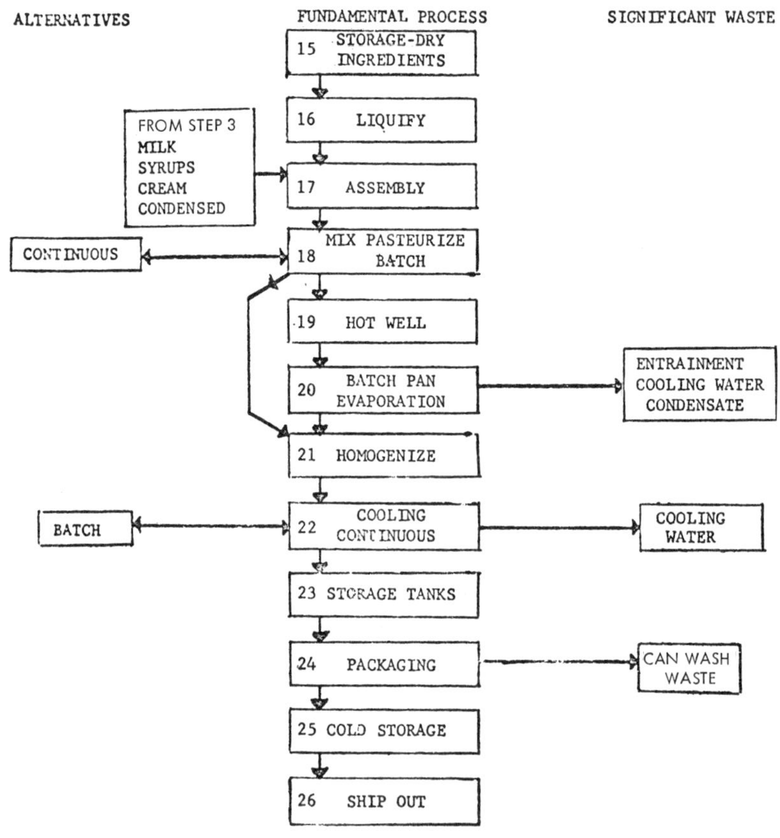

Source: FWPCA Publication IWP-9

1. Receipt — Dairy products are normally received in tank trucks and ten-gallon cans. The larger volume plants tend to receive in tank trucks and the smaller plants in cans. Larger plants may receive a small portion in cans. Dairy products normally received include milk as well as skim milk, whey, and buttermilk, which are by-products of other processing operations. Another significant amount of receipts is in the form of liquid sugar and

liquid corn syrup for use as mix ingredients (although smaller operations may receive the sugar in bag form). The liquid ingredients are pumped to storage tanks.

2. Raw (Unpasteurized) Product Storage — Dairy products are stored in refrigerated tanks prior to utilization. The sugars and corn syrups are stored in heated tanks prior to use.

3. Separation — From storage a portion of the dairy products is sent through a heating device and then to a centrifugal separator in which all of the cream is separated and sent to storage for later use in ice cream mix manufacture. The skim milk is sent to pasteurization equipment.

4. Pasteurization — The milk is usually pasteurized in a continuous flow pasteurizer, although smaller plants may continue to use vat-type pasteurizers. The pasteurized milk is sent to a surge tank to adjust the difference between flow rates of the pasteurizer and of the evaporator.

5. Evaporation (Condensing) — Water is evaporated from the milk by heating the milk with steam in a vacuum chamber. A vacuum is maintained so that the boiling point is lowered to a point at which the product is not injured through excessive heat. The evaporator is normally a continuous flow unit, although in smaller operations batch type evaporators may be used. The continuous evaporator may have a number of "effects" which permit a significant reduction in the amount of steam necessary to produce a given amount of evaporated milk.

Since the first cost of adding "effects" is significant, the tendency to have more effects occurs in the larger plants where greater economies are realized. Normally 15% of the water is removed and the evaporated milk will have 25.5% solids compared to the original milk, 12.5% solids. The evaporated (condensed) milk may be sent to (12) spray dryers or may be sent to cooling.

6. Cooling — Product from the evaporator is usually cooled in a continuous cooler, although occasionally batch cooling is used.

7. Pasteurized Storage — The cooled evaporated milk is stored in refrigerated trucks.

8. Packaging in Cans — Automatic machinery is used to fill and seal the cans.

9. Sterilize — The canned milk proceeds to the sterilizer where it is heated to a high temperature for a short period of time and then cooled to storage temperature.

10. Storage — Packaged product is inventoried in storage until needed for delivery to customers.

11. Ship Out — The finished product is drawn from storage and placed on vehicles for distribution.

12. Spray Drying — That portion of the evaporated milk to be used for powder is sent to a spray dryer. The product is pumped with a high pressure pump into a heated air screen where the remaining moisture is evaporated and the dry powder is separated. The spray dryer is a continuous form of drying under relatively low heat contact with the product. Some evaporated milk continues to be dried on roll dryers.

13. Instantizing — Powdered milk used by the consumer is usually sent through an instantizing process prior to packaging.

14. Packaging Powder — The powdered milk from the dryer or from the instantizer is packaged in bags and barrels for bulk users and in retail type packages for consumer use. The finished product is sent to Storage (10).

In addition to the evaporated milk and powdered milk process described, there is the additional fundamental process of mix-making in this type of establishment. From Step 3 above, cream is available from refrigerated storage tanks. From Step 1, dairy products and sugars.

15. Storage of Dry Ingredients — In addition to the liquid dairy and syrup ingredients, dry ingredients such as stabilizers, emulsifiers, and chocolate are needed, as well as water. The dry ingredients are held in storage prior to use.

16. Liquefy Dry Ingredients — Prior to use, the ingredients are placed into solution in a mixing device and then pumped to assembly.

17. Assembly — All of the ingredients listed above are assembled in mixing tanks in the correct proportions needed for the final "mix".

18. Mix Pasteurization — The assembled mixes are pasteurized in batch quantities, although in very large plants a continuous flow pasteurizer may be used.

19. Hot Well — Those mixes that contain excess water are pumped to the hot well (which is a surge tank) and from there pumped to the evaporator.

20. Batch Evaporating Pan — Excess water is removed [as described in (5)] as required by the mix formulation and will be pumped to the homogenizer.

21. Homogenizer — Mix from the batch pan or from the continuous pasteurizer is pumped to the homogenizer, a high pressure pump which breaks up fat particles within the mix to insure that they stay in suspension in the finished product.

22. Cooling — From the homogenizer the warm mix is sent through a continuous cooler, although in small plants a batch-type cooler may be used.

23. Pasteurized Storage — The cooled mix is stored in refrigerated tanks until ready for further use.

24. Packaging — Generally, the finished mix is packaged in 10-gallon cans and to a lesser extent in single service plastic bag and cardboard box type containers. Occasionally large volumes of mix are shipped out in tank truck quantities to the customer.

25. Cold Storage — The canned or packaged mix is held in refrigerated storage as inventory before shipping.

26. Shipping Out — The packaged mix is drawn from cold storage and placed on refrigerated trucks for delivery to the customer.

The fundamental condensed and evaporated milk processes have changed little in recent years, and little change is forecast to 1977. Nevertheless, several developments of interest have occurred. The trends may best be shown in the table below. The reader should note that the alternative subprocesses and other industry changes have occurred over a span of years. The process which will become prevalent is identified as (P), and that which is becoming less used as (S). The estimates represent the observations and opinions of people in the industry.

TABLE 12: ESTIMATED PERCENTAGE OF PLANTS EMPLOYING PROCESS

		1950	1963	1967	1972	1977
P	Receive in Tank Trucks	-0-	40	50	60	70
S	Receive in Cans	100	60	50	40	30
P	Pasteurize Continuously	25	50	60	70	90
S	Pasteurize Batch	75	50	40	30	10
P	Spray Dry	50	75	85	90	95
S	Roll Dry	50	25	15	10	5
P	Automatic Packaging	-0-	20	25	35	45
S	Manual Packaging	100	80	75	65	55
P	CIP Piping	-0-	40	50	60	70
S	Take-apart Piping	100	60	50	40	30
P	Automatic Material Handling	-0-	50	60	70	80
S	Manual Material Handling	100	50	40	30	20

Source: FWPCA Publication IWP-9

The subprocesses shown in Table 12 do not require different treatment from the fundamental processes; however, the choice of subprocess is largely determined by the total volume produced. The continuous flow processes tend to have less waste per pound of finished product because of the greater productivity per piece of equipment. Product spillage and waste during normal processing, and cleaning water and soaps represent the significant wastes for any type process used.

In order to best estimate total industry waste and wastewater, it is desirable to identify levels of technology within the industry. The following table illustrates three technological levels. The fundamental process steps from Figure 8 are used as reference for the table which follows.

TABLE 13: COMPARATIVE TECHNOLOGY

Older Technology	Typical Technology	Advanced Technology
1. Receive product in 10-gallon cans	Receive bulk of dairy products in tank trucks and part of products in 10-gallon cans	Receive all products in tank trucks
2. Store product in 10-gallon cans	Products stored in tanks	Products stored in refrigerated tanks
3. Heat and separate centrifugally	Heat and separate centrifugally	Heat and separate centrifugally
4. Pasteurize in batch quantities	Pasteurize in a continuous unit and pump to (5)	Pasteurization on a continuous basis
5. Evaporate (condense) in batch quantities	A continuous evaporator (condenser) which will have one or two effects	Evaporation (condensing) on a continuous basis through a double or triple effect pan
6. Cooling in batch quantities	Cooling in a continuous process	Cooling in a continuous process
7. Pasteurized storage in 10-gallon cans	Pasteurized storage in tanks	Refrigerated pasteurized storage
8. Packaging in cans	Packaging automatically in cans	Packaging performed automatically in high-speed can machinery
9. Sterilization of canned products in batch equipment	Cans sterilized in a continuous unit	Sterilization in a continuous sterilizer
10. Storage inventoried finished product in warehouse	Inventory finished product in storage room	Finished product stored in palletized quantities and (11)
11. Ship out	Ship out	Ship out in pallet loads
12. Drying performed on a "drum dryer"	Drying performed in a spray type unit	Spray drying utilized
13. Instantizing not used	Powder conveyed to an instantizer and then conveyed to (14)	Powder conveyed to an instantizer and in turn conveyed to (14)
14. Packaging powdered product manually	Automatic packaging machinery	Automatic packaging machinery. Packages automatically boxed and automatically palletized
15. Storage of mix ingredients in dry quantities in bag form	Storage of dry ingredients for mix, a minimum of bag ingredients and as many as possible in liquid or syrup form	Storage of mix ingredients all in liquid form in tanks

(continued)

TABLE 13:　(continued)

Older Technology	Typical Technology	Advanced Technology
16. Liquefy dry ingredients manually	Liquefication of dry ingredients performed mechanically	Liquefication of dry ingredients no longer necessary
17. Assemble directly into (18)	Assembly takes place in a tank on scales with ingredients measured manually	Assembly takes place in programmed automatic vat-on-weigh scales
18. Batch pasteurizers	Mix pasteurization centrifugally in batch quantities which are pumped to to (19)	A continuous mix pasteurizer which proceeds to (19)
19. Pump to hot well and then pumped to (20)	A hot well and from there pumped to (20)	A hot well and (20)
20. The batch evaporating pan	The batch evaporating pan	A continuous pan
21. Homogenize and pressure pump	Homogenization by pressure pump	Homogenization by pressure pump
22. Cool in batch quantities	Cooling in a continuous cooler	Cooling in a continuous manner
23. Store pasteurized product in 10-gallon cans, which will be the same cans for (24)	Pasteurized storage in refrigerated tanks	Pasteurized product stored in refrigerated tanks
24. Packaging	Packaging of mix partially in 10-gallon cans and partially in single service bag-in-box containers	Packaging in single service containers such as bag-in-box
25. Store packaged cans in cold storage	Finished product inventoried in cold storage	Inventoried in cold storage in palletized quantities
26. Shipped out	Shipped out	Shipped out in palletized lots
27. Take-apart piping	Partial CIP piping	CIP piping
28. Manual material handling	Partial automatic material handling	Automatic material handling

Source:　FWPCA Publication IWP-9

WASTE PRODUCTION BY THE INDUSTRY

Dairy plant wastes consist mainly of lost raw materials, intermediate and finished products and the cleaning materials required to clean and sanitize shipping containers and processing space and equipment all carried in the process waters being discharged by the plant. In addition, whey is a by-product of most cheese operations and can become a significant pollution problem. Every effort should be made to keep it out of the sewer system when it can readily be separated from the water being used in the plant. The usual procedure is to concentrate the whey so it can be dried to whey powder or converted into another usable byproduct either at the plant produced or at another location which can economically be reached from the point of production. These whey byproducts are then used as food or feed supplements.

Milk plant wastes normally have a BOD concentration ranging roughly between 500 and 4,000 mg./l. and a COD concentration which usually runs 2 to 2.5 times higher, according to Watson (13). As a general rule, the suspended solids in such wastes are not high enough to be of great significance. Cleaning compounds being used, particularly in larger plants, can be responsible for swings of consequence in the pH of the effluent but these can normally be satisfactorily corrected by equalization by the city or the plant generating the wastes.

Prior to quite recently, the majority of available information in respect to dairy food plant wastewater composition was limited to BOD, suspended solids and COD. Very limited information on dairy food plant wastewater is available for temperature, pH, fat, protein, carbohydrates and other components such as phosphates, chloride, sulfur, surfactants and sanitizers.

BOD VALUES

The biological and chemical oxygen demand of milk plant wastewater will vary as a function of the products manufactured since differences occur in the amount of oxygen that is required for the oxidation of different constituents such as fats, carbohydrates and protein. Various dairy products differ widely in their relative concentrations of major organic constituents and an understanding of the composition of various fluid dairy foods is essential to a full understanding of dairy plant wastes.

The constituents of primary concern in water pollution for various dairy products are cited in Table 14. These values are based upon current average compositional values and vary somewhat in previously cited compositional data for dairy products in early waste guides (3).

Reported BOD values for various milk constituents and other organic components that may be present in dairy food plant wastewaters are listed in Table 15. The major three constituents of milk contributing to BOD are lactose, milk fat and milk proteins with BOD_5/lb. of component averaging 0.65, 0.89 and 1.03 respectively. Utilization of these figures permits the calculation of BOD and the relative contribution to BOD by various constituents.

Calculated values, literature values and percentage contribution of milk components to BOD in different products are cited in Table 16 for most common dairy foods. Except for whey, the calculated and literature values are in close agreement.

Frequency plots of the variability in day to day BOD_5 values for six different dairy plants are shown in Figure 9. Five of the six plots follow a normal distribution and show a straight-line relationship, whereas the sixth plot shows an S-shaped distribution typical of very poor day to day control.

Where normal distribution existed, the maximum BOD level (at a 1% confidence limit) ranged from 1.77 to 2.46 times the mean value, with an average of 2.1. For the plant with abnormal distribution, the maximum BOD was 3.6 times the mean value.

Variation is a characteristic of dairy food plant wastes which needs to be evaluated on a plant to plant basis in order to provide information for both design and proper operation of waste treatment systems. On a daily basis, control of variables in the proper process can result in a normal and predictable distribution around the mean. Generally, maximum values at 1% confidence limit can be predicted by multiplying the mean value by 2.5.

TABLE 14: AVERAGE COMPOSITION OF MILK AND MILK PRODUCTS (100 GRAMS)

Product	Fat g.	Protein g.	Lactose g.	Lactic Acid g.	Total Organic Solids g.	Ca mg.	P mg.	Cl mg.	S mg.	Total Ash g.	Viscosity (c.p. at 20°C)
Skimmilk	0.08	3.5	5.0	–	8.56	121	95	100	17	0.7	1.4
2% milk	2.0	4.2	6.0	–	12.2	143	112	115	20	0.8	2.4
Whole milk	3.5	3.5	4.9	–	11.1	118	93	102	19	0.7	2.2
Half & half	11.7	3.2	4.6	–	19.5	108	85	90	16	0.6	7.5
Coffee Cream	18.0	3.0	4.3	–	25.3	102	80	73	12	0.6	15.0
Heavy cream	40.0	2.2	3.1	–	45.3	75	59	38	9	0.4	25.2
Choc. milk[a]	3.5	3.4	5.0	–	18.5	111	94	100	19	0.7	15.0
Churned Buttermilk	0.3	3.0	4.6	0.1	8.0	121	95	103	15	0.8	1.5
Cultured Buttermilk	0.1	3.6	4.3	0.8	10	121	95	105	17	0.7	500
Sour cream	18.0	3.0	3.6	0.75	24.6	102	80	73	12	0.6	10000
Yoghurt[b]	3.0	3.5	4.0	1.1	10.8	143	112	105	19	0.7	3000
Evap. milk	8.0	7.0	9.7	–	27	757	205	210	39	1.6	30
Ice cream[c]	10.0	4.5	6.8	–	41.3	146	115	104	20	0.9	35.0
Whey	0.3	0.9	4.9	0.2	6.3	51	53	195	8	0.6	1.4
Cot. cheese Whey	0.08	0.9	4.4	6.7	6.1	96	16	95	8	0.8	1.3

Organic Ingredients Added
 a. Sucrose, 6% & chocolate solids, 1%

 b. Fruits

 c. Sugar, 15%

Source: Second National Symposium on Food Processing Wastes (1971)

TABLE 15:   REPORTED BOD5 VALUE FOR VARIOUS MILK CONSTITUENTS

| | Pound BOD_5 / Pound Component | |
Constituent	Range	Average
Lactose	.42 – .72	.65
Glucose	.53 – .78	.66
Lactic Acid	.63 – .64	.63
Milk Fat	.70 – .95	.89
Glycerol	.65 – .83	.75
Butyric Acid	.34 – .90	.75
Sodium Butyrate	–	.41
Palmitic Acid	–	1.07
Sodium Palmitate	.4 – 1.02	.70
Stearic Acid	–	.80
Sodium Stearate	.45 – 1.70	1.20
Milk Proteins	.60 – 1.20	1.03
Casein	.25 – 1.17	1.04
Hydrocolloids		
Sodium Alginate	–	.36
Carboxymethyl-cellulose	–	.30

TABLE 16:   LITERATURE AND CALCULATED VALUES AND CONTRIBUTION OF MILK COMPONENTS TO BOD5

| | | | % Contribution to BOD_5 by | | | |
Product	Literature BOD_5	Calculated BOD_5	Milk Fat	Milk Protein	Lactose	Lactic Acid
Skimmilk	67,000	73,000	6.3	49.3	44.5	–
2% milk		100,000	17.8	43.3	39.0	–
Whole milk	104,000	99,000	31.5	36.4	32.2	–
Half and half	156,000	167,000	62.4	19.7	17.9	–
Coffee cream	206,000	219,000	73.1	14.1	12.8	–
Heavy cream	399,000	399,000	89.2	5.7	5.0	–
Choc milk	–	145,000	21.5	24.2	22.4	–
Churned buttermilk	68,000	71,000	4.2	48.2	46.7	0.98
Cultured buttermilk	–	64,000	1.3	52.2	39.4	7.1
Sour Cream	–	218,000	73.5	1.4	10.7	21.7
Yoghurt	–	91,000	27.8	37.6	27.1	7.2
Evaporated milk	208,000	206,000	34.6	35.0	30.6	–
Ice Cream	292,000	290,000	30.7	15.9	15.2	–
Whey	34,000	45,000	5.9	20.6	70.8	2.8
Cottage cheese whey	31,500	42,000	1.6	22.1	68.1	10.5

Source: Second National Symposium on Food Processing Wastes (1971)

FIGURE 9:   DAY TO DAY VARIATIONS IN BOD IN THE WASTEWATER
OF DIFFERENT DAIRY FOOD PLANTS

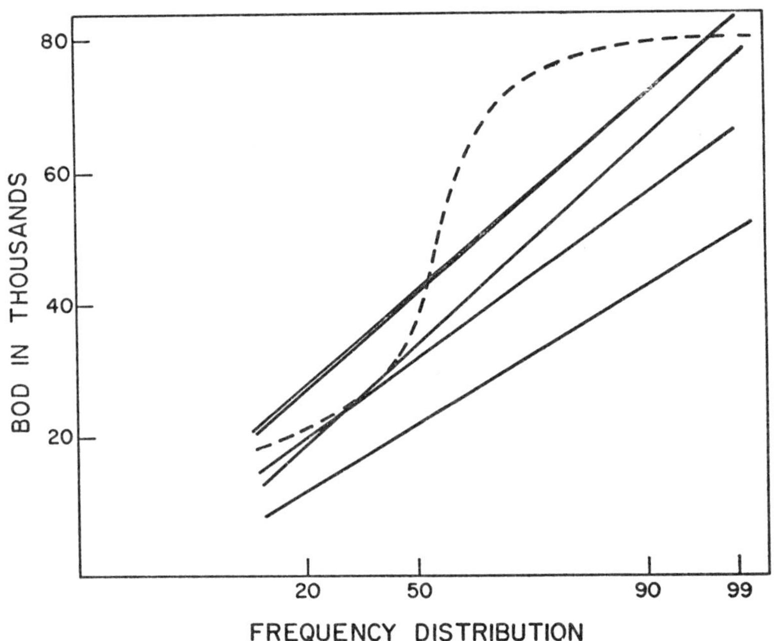

Source: Second National Symposium on Food Processing Wastes (1971)

Table 17 is the average reported BOD coefficients (pound BOD_5/1,000
pounds of milk produced) for various fluid dairy plant operations reported
in the literature. The earliest reported data were those of Trebler and
Harding (16) in 1947, which were revised in 1949, and subsequently revised by a dairy industry committee and published in the 1967 issue of
Milk Industry Foundation Milk Plant Operator's Manual.

The Scottish Milk Market Board and Royal in England have also presented
data on BOD coefficients. Although the values do not cover exactly the
same unit operations, the coefficients for unit processes are in good agreement with the revised Trebler and Harding data and can be utilized to
develop an indication of waste loads in different types of combination
plants. It should be emphasized that, whereas whey is the most visible
BOD source (260 pounds BOD/1,000 pounds milk processed), other dairy
food plant operations are significant sources of BOD.

TABLE 17: AVERAGE REPORTED BOD LOSSES FOR VARIOUS DAIRY FOOD PLANT UNIT OPERATIONS

Process	Pounds BOD/1000 Pounds Milk Processed				
	Trebler & Harding, 1947	Trebler & Harding, 1949	MIF Manual, 1967	Scotland 1968	Royal 1968
Receiving and cooling milk	0.6	0.4	0.2	0.25	0.65
Tanker delivery, washing	0.1	0.1	0.1	0.25	--
Storage of milk in tanks	0.1	0.1	0.05	0.18	--
Cream separating	0.4	0.2	0.2	0.13	--
Cream pasteurization, cooling, filling	0.4	0.2	0.2	0.75	0.96*
Cream separation and pasteurization	--	--	--	0.65	--
Skim milk past, cooling, can filling	--	0.2	0.2	--	--
Milk pasteurization and storage	--	--	--	0.29	--
Milk past., storage and bottling (glass)	0.8	0.8	0.8	0.85	1.08
Milk past., storage and bottling (paper)	--	0.6	0.6	--	--
Bottling pasteurized milk	--	--	--	0.11	--
Glass bottle washing	--	--	--	0.23	--
Cottage cheese (casein) mfg.	1.6	1.6	4.7	15.0	--
Cheese (pressed) mfg. (unwashed curd)	--	--	0.8	--	0.89
Evaporating whole milk					
Floor waste	0.3	0.2	0.2	--	--
Entrainment losses	0.2	0.1	0.1	--	--
Evaporating skim milk	--	--	--	0.23	--
Canning sterilizing evap. milk	0.3	0.2	0.2	--	1.08**
Condensing whey (2:1)	--	--	--	0.29	--
Condensing whey (3:1)·	--	--	--	1.40	1.92
Sweet whey					
Floor wastes	0.8	0.8	0.8	--	--
Condensate	0.4	0.4	0.4	0.75	--
Washing condenser	--	--	--	0.60	--
Acid whey					
Floor wastes	0.8	0.8	0.8	--	--
Condensate	0.4	0.3	0.3	0.60	--

(continued)

TABLE 17: (continued)

Process	Pounds BOD/1000 Pounds Milk Processed				
	Trebler & Harding, 1947	Trebler & Harding, 1949	MIF Manual, 1967	Scotland 1968	Royal 1968
Condensing sweetened condensed skim milk	--	--	--	1.40	--
Floor wastes	0.6	0.3	0.3	--	--
Entrainment loss	0.3	0.05	0.05	--	--
Condensing & canning sweetened milk	--	--	--	--	1.32
Supercondensing skim	1.2	1.2	1.2	--	--
Spray drying whole milk	0.2	0.1	0.1	--	--
Roller drying	--	--	--	0.51	1.80
Evaporating skim and drying	--	--	--	0.74	1.02
Whey drying	0.6	0.5	0.5	--	--
Ice cream making and filling	--	--	--	1.60	--
Ice cream making, pan	0.6	0.4	0.4	--	--
Ice cream making, vat	0.6	0.4	0.4	--	--
Ice cream freezing	0.2	0.05	0.05	--	--
Cultured buttermilk making	1.0	0.5	0.5	--	--
Sour cream mfg.	--	--	--	1.2	--
Butter churning and washing	0.6	0.2	0.2	0.46	0.72

*Includes separation
**Includes condensing

Source: EPA Report 12060 EGU, March 1971

During the course of this investigation, data was obtained from two different modern automated dairy plants, both producing milk and cottage cheese, as shown in Table 18.

TABLE 18:   INDUSTRIAL DATA FOR BOD COEFFICIENTS FOR UNIT PROCESSES IN TWO FLUID MILK, COTTAGE CHEESE PLANTS

Process	Pounds of BOD/1000 Pounds of Milk Received	
	Plant A	Plant B
Tank truck rinsings	0.25	--
Silo tank rinsings	0.23	--
Tank trucks and storage tank rinsings	--	0.25
Separation (CIP)	0.15	--
CIP Separator sludge	--	0.08
HTST start-up and change-over and cleaning	0.75*	0.55
Filling operations-milk	0.30	0.32
Lubricants	--	0.08
Product returns	--	0.10
Cottage cheese wash water	0.33	0.60
Unaccounted	0.20	0.03
TOTAL WASTE WATER	2.21	2.01

*Initial start-up collected, product change-overs not diverted

Source: EPA Report 12060 EGU, March 1971

WASTEWATER VOLUMES

Water usage for dairy processing in the United States is greater than 27 billion gallons (3). The increasing use of user charges in the form of surcharges are affecting the dairy processing industry. Increasing costs of water and waste treatment in an industry with a small profit margin make

careful management of water and wastewater necessary. Limited results
have indicated that a 50% reduction in water use in a small-size dairy
plant is possible with an employee motivation program (8). Also, BOD
was reduced by 50% with changes in processing procedures. Water usage
is approximately 25 times higher in "Older Technology" than for "Advanced
Technology" plants and BOD generated is 5 times as great for the former (3).

TABLE 19:   COEFFICIENT WATER USE FOR VARIOUS PLANT OPERATIONS

Operation	Pounds of Water/Pounds of Milk Processed
Can washing	0.001
Truck washing	0.07
Cooling milk in storage	0.2
Pasteurization	0.35
Bottle washer	0.5
Boilers & Evaporators	0.25
Cleaning	0.25
Spray cleaning tanks	
Pre-rinses	0.005
Cleaning	0.010
Post-rinse	0.005
Sanitizing	0.005
Lines	
Pre-rinse	0.0005
Cleaning	0.001
Post-rinse	0.0005
Sanitizing	0.025
Evaporating (3:1) (triple effect) Calander cond.	9.0
Condensing (triple effect) Plate cond.	12.0
Cottage cheese wash water	1.2
Cooling	
Receiving, bottling, ice cream & cream cheese	1.2
Receiving past. separating, condensing & spray drying	1.1 - 1.5
Receiving cond. canning & sterilizing	0.02- 0.4

Source: EPA Report 12060 EGU, March 1971

Relatively little information is available concerning the amount of water per unit weight of milk processed as utilized in various types of dairy food plant operations. Information available in the literature and that obtained from industry has been summarized and presented in Table 19. Major processes utilizing water are washwater for cottage cheese, cooling water, and condensing water. Cooling waters are generally free from waste and could be segregated from other operations to minimize hydraulic loading up to 70%.

The strength of dairy food plant wastewaters has been reported generally in terms of parts per million of BOD, without regard to total wastewater volume or the relationship of the amount of milk processed; and waste volumes are often reported irrespective of the volume of product processed. Wastewater volume and strength coefficients have been calculated from industrial and literature data in respect to the amount of milk, or milk processed. For the purpose of the present discussion, the wastewater volume coefficient (WWVC) expressed in pounds of wastewater per pound of milk processed and the wastewater strength coefficient (WWSC) expressed in pounds of BOD_5 per 1,000 pounds of milk processed are defined as follows:

$$WWVC = \frac{\text{gal. wastewater} \times 8.34}{\text{gal. milk processed} \times 8.6}$$

$$WWSC = \frac{\text{lbs. } BOD_5 \text{ per day}}{\text{lbs. milk processed per day} \times 1,000}$$

A summation of literature values for wastewater coefficients for various types of dairy food plant operations are shown in Table 20. The use of volume as a reference for the coefficients eliminated the "yield" variable for different products and brings the coefficients into relatively close agreements for different types of dairy food plants.

Through personal contacts and plant site visitation, information on dairy plant wastes and waste treatment was obtained for 57 plants in terms of milk volume processed, gallons of wastewater produced, and ppm BOD_5. The pounds of BOD per day, volume coefficient and BOD coefficients were calculated (Table 21). All of the plants surveyed used either advanced technology or a mixture of advanced and typical technology with 30 of the 57 plants manufacturing more than one type of dairy product.

Statistical analyses indicated no significant difference in BOD or volume coefficients and the type of plant operation except that BOD coefficients of the plants processing milk plus cottage cheese and ice cream had significantly higher coefficients than the other type of operation. Coefficients were unrelated to plant size or to the degree of automation. Plants that were fully automated in respect to processing and CIP systems frequently

had higher than average volume and BOD coefficients. The wide variability of both volume and BOD coefficients among plants of similar technologies and size and a lack of correlation between waste coefficients
suggests that the controlling factor in waste volume and BOD coefficients
was management related.

TABLE 20: SUMMARY OF LITERATURE VALUES OF WASTEWATER COEFFICIENTS*

Products Mfg.	No. of Plants	Waste Volume Coef.		BOD_5 Coef.	
		Range	Average	Range	Average
Milk Receiving	6	4.6 - 12.5	6.1	0.02 - 4.8	1.0
Milk	11	1.5 - 18.6	4.9	1.1 - 22.0	5.2
Butter	2	1.4 - 8.3	4.85	0.8 - 2.1	1.46
Cheese	12	0.3 - 5.1	2.06	0.2 - 4.1	1.8
Condensed	2	1.2 - 2.3	1.75	1.0 - 1.9	1.45
Powder	10	0.8 - 11.5	2.8	0.6 - 12.3	3.9
M. CC. IC.	2	0.8 - 1.2	1.0	0.6 - 0.9	0.7
Mix Prod.	5	1.1 - 6.8	1.2	1.3 - 3.2	1.9
	46	0.3 - 18.6	3.9	0.2 - 22.0	2.6

* Volume coefficient in #/# milk processed, BOD_5 coefficient in $\#BOD_5$/1000
#milk or milk equivalent received.

Source: Second National Symposium on Food Processing Wastes (1971)

TABLE 21: SUMMARY OF COMMERCIAL PLANT SURVEY FOR WASTE-WATER COEFFICIENTS*

Products Mfg.	No. of Plants	Waste Vol. Coef.		BOD Coef.	
		Range	Average	Range	Average
Milk	6	0.1 - 5.4	3.25	0.2 - 7.8	4.2
Cheese	3	1.63 - 5.7	3.14	1.0 - 3.5	2.04
Ice Cream	6	0.8 - 5.6	2.8	1.9 - 20.4	5.76

(continued)

TABLE 21: (continued)

Products Mfg.	No. of Plants	Waste Vol. Coef. Range	Average	BOD Coef. Range	Average
Cond. Milk	2	1.0 - 3.3	2.1	0.2 - 13.3	7.6
Butter	1	-	0.8	-	0.85
Powder	2	1.5 - 5.9	3.7	0.02 - 4.6	2.27
Cottage Cheese	3	0.8 - 12.4	6.0	1.3 - 71.2[+]	34.0
Cottage Cheese & Milk	19	0.05 - 7.2	1.84	0.7 - 8.6[+]	3.47
Cottage Cheese Ice Cream & Milk	9	1.4 - 3.9	2.52	2.3 - 12.9	6.37
Mixed Products	5	0.8 - 4.6	2.34	0.9 - 6.95	3.09

*See footnote, Table 20
+Whey included, whey excluded from all other operations manufacturing cottage cheese.

Source: Second National Symposium on Food Processing Wastes (1971)

BOD-COD INTERRELATIONSHIPS

The 1959 Revised Guide for Dairy Plant Waste Treatment (11) recommended
the utilization of COD in preference to BOD_5 as a means of measuring the
strength of the wastes. This recommendation was based on the work of
Porges and co-workers (17) utilizing skim milk as a model for dairy food
plant wastes. Investigators are in agreement in respect to the BOD-COD
relationships between whole milk, skim milk and whey. The average values
cited are:

Product	BOD-COD Ratio
Whole milk	0.69
Skim milk	0.63
Buttermilk (churned)	0.66
Whey	0.52

(continued)

Product	BOD-COD Ratio
Lactose	0.53
Casein	0.46
Whey protein	0.23
Fat	1.28

With the exception of fat, the BOD-COD ratios of the constituents of milk are much lower and the lowest biologically oxidizable material is the whey protein.

BOD-COD ratios reported in the literature for industrial dairy food plant wastes are shown in Table 22. Values vary widely and are generally less than for milk products themselves, as might be anticipated. Overall, the BOD-COD ratio ranged from 0.10 to 0.88 with an average of 0.53, as compared to a BOD-COD ratio of 0.65 for whole milk.

Values below 0.6 can be interpreted to suggest less efficient biological oxidation of milk wastes than pure milk, probably caused by the presence of nonmilk constituents. Possible "toxicity" of dairy plant waste is suggested by values below 0.4, but no reference to possible "toxicity" has been previously cited.

In the course of the investigation reported by Harper et al (15), and/or regional dairy companies who have had a long experience in measuring the strength of dairy plant wastes, indicated they had attempted to utilize COD but abandoned the position several years ago because of wide variations in BOD-COD ratios and apparent lack of agreement between values suggesting that COD values were less reproducible and of less value than BOD.

Data was obtained from one plant for BOD-COD ratios in raw and treated wastes at hourly intervals during a production day. As shown in Table 23, the ratios varied from 0.12 to 0.90 at different periods of the day. Low ratios between 1:00 and 5:00 P.M. coincided with the major periods of equipment process cleaning. These low values suggest either "toxicity" from detergents or the presence of a large amount of refractory material.

TABLE 22:   BOD$_5$/COD RATIOS FOR RAW DAIRY PLANT WASTES

Investigator	Type of Plant	Year	BOD$_5$/COD ratios Range	Average
Hoover & Porges	–	1953	0.34-0.80	–
Schulz-Fulkenheum	–	1955	0.30-0.70	–

(continued)

TABLE 22:   (continued)

Investigator	Type of Plant	Year	BOD$_5$/COD ratios Range	Average
Walholz	–	1956	–	0.64
Furoff	multi-produce	1960	0.22–0.51	0.33
Schweizer	cheese	1960	0.31–0.66	0.45
Rensink	butter	1962	–	0.66
Christansen	–	1964	0.55–0.77	0.64
Walgren	–	1966	0.11–0.75	0.47
Bergman, et al	multi-product	1966	0.43–0.60	0.53
Bergman, et al	multi-product	1968	0.40–0.75	0.56
Annon	–	1968	0.22–0.88	0.35
Current plant survey	–		0.10–0.76	0.48

TABLE 23:   BOD/COD RELATIONSHIPS IN RAW AND TREATED DAIRY PLANT WASTES (1967) AT HOURLY INTERVALS

Time	BOD:COD Ratios Raw	Treated	Rate of Waste Water Flow (mgd)	Raw COD Values
8:00 a.m.	.4	.32	.025	3440
9:00 a.m.	.35	.29	.020	3200
10:00 a.m.	.51	.26	.025	3400
11:00 a.m.	.28	.25	.010	4700
12:00	.66	.15	.035	2600
1:00 p.m.	.26	.48	.035	3400
2:00 p.m.	.21	.90	.035	5200
3:00 p.m.	.21	.82	.050	5800
4:00 p.m.	.12	.77	.035	9280
5:00 p.m.	.25	.59	.045	6240
6:00 p.m.	.45	.75	.035	7760
7:00 p.m.	.60	.57	.014	3200
8:00 p.m.	.57	.68	.020	4480
9:00 p.m.	.55	.54	.020	2720
10:00 p.m.	.88	.64	.022	2400
11:00 p.m.	.90	.63	.010	1056
12:00	.90	.60	.005	886

*Equipment sanitized 6-8 A.M. Cleanup 1-6 P.M., maximum activity 2-4 P.M.

Source: Second National Symposium on Food Processing Wastes (1971)

SOLIDS, pH AND TEMPERATURE

Available information from industry for solids, pH and temperature of dairy
food plant wastewater in relation to BOD strength is summarized in Table
24 by Harper et al (15). The suspended solids content of dairy plant wastewaters from the literature varied between 2,400 and 4,500 ppm. Based
on the limited literature values available, there was no statistically significant correlation between the suspended solid strength and the type of
dairy plant operation.

The data suggest that over 70% of the suspended solid is volatile, or organic
in nature, in dairy food plant wastewaters. Volatile suspended solids, as
percentage of total suspended solids, ranged from 68 to 98% with an average of 85%. The suspended solids are composed primarily of coagulated
milk, fine particles of cheese curd and pieces of fruits and nuts from ice
cream operation, and contribute from 10 to 30% of the total solids. Volatile solids as percentage of total solids, ranged from 36 to 96% with an
average of 63%.

The nonorganic total solids is about twice the nonorganic volatile suspended
solids. The proportion of nonorganic solids in milk is about 13%, which
suggests that most of the suspended solids in dairy food plant wastewaters
are of a dairy food origin. The increase of nonorganic matter in a total
solid reflects the contribution of nonorganic material from detergents,
sanitizers and lubricants.

The pH of the dairy food plant wastewaters varied from 4.4 to 9.2 with
a median value of 7.2. Although it would be expected that cheese plants
might have a lower pH in their wastewater than other types of plants, this
was not found in the data available. The major factor affecting pH of dairy
food plant wastes is the cleaning compound, either acid or alkali and its
relative significance in the dairy food plant wastewater.

Suspended solids to BOD ratios ranged from 0.25 to 0.35 with an average
of 0.31, and the total solids to BOD ratios were essentially identical for
different types of dairy plant wastes with an average of 1.62. The relationship of solids to BOD appeared to be independent of the type of dairy
plant operation and characteristic of all dairy plant wastewaters.

Only limited studies have been made of the temperature of dairy food plant
wastewaters as they leave the dairy plant. Reported values ranged from
70° to 115°F. (20° to 40°C.). The most comprehensive study in this area
was made by Zall (8) who showed that wastewater temperature in a small
dairy plant could be reduced about 15% by management control of the use

of hot water in cleaning, and by eliminating the practice of allowing hot water hoses to run while not in actual use.

TABLE 24: SOLIDS, pH AND TEMPERATURE OF DAIRY FOOD PLANT WASTEWATERS (INDUSTRY VALUES)

Waste Component	No. of Values	Range	Average
Suspended Solids, ppm	24	24-5700	-
Volatile Suspended Solids, ppm	19	17-5260	-
Volatile Total Solids, ppm	22	57-4700	1497
Total Solids, ppm	27	135-8500	2397
pH	30	5.3-9.4	7.1
Temperature, °F	17	55-120	76
BOD, ppm	11	15-4790	2100

Source: Second National Symposium on Food Processing Wastes (1971)

CHEMICAL COMPOSITION

No systematic complete study of the chemical composition of dairy plant wastewaters has been made to date, and only limited data is available. The most comprehensive studies have been those of Walgren (17A) and Sarkka et al (18), for dairy plants in Sweden and Finland, respectively. Compositional data for the major organic constituents in dairy food plant wastes are essential to a better understanding of the consistency of the composition of dairy wastes and the oxygen requirements of these waste-waters for treatment, since the oxygen consumed varies from compound to compound.

Information on mineral constituents of dairy food plant wastes is important because of the role of the various refractory elements in water quality. The organic composition and concentration of phosphate and chloride reported in various studies are presented in Table 25 for literature and industrial survey data. Limited data indicate the following differences in the ratios of fat to protein and to lactose.

Source	Protein/Lactose Ratio	Fat/Lactose Ratio
Whole milk	0.64	0.70
Dairy wastewater	0.56	0.12
Dairy wastewater	0.43	----

Source	Protein/Lactose Ratio	Fat/Lactose Ratio
Dairy wastewater	0.85	0.57
Dairy wastewater	1.08	0.42
Dairy wastewater	1.14	----
Dairy wastewater	0.32	0.05

Protein/lactose ratios and fat/lactose ratios in dairy food plant wastes are
significantly different from those for whole milk. The phosphate values
which vary independently of BOD, were all in excess of 10 ppm and could
be considered as a potential source of phosphate for enhancing algae growth.
The values varied widely, apparently reflecting a difference in the amount
of phosphate cleaning compound used in different dairy operations. Most
chloride values in wastewater exceeded 75 ppm which is considered to be
the level that cannot interfere with industrial processes. Forty-five percent
of the chloride values given exceeded 250 ppm.

TABLE 25:   CHEMICAL COMPOSITION OF DAIRY FOOD PLANT WASTE-
WATER

| | PPM | |
Constituent	Range	Average
BOD	450 – 4,790	1,885
Nitrogen	15 – 180	76
Protein	210 – 560	350
Fat	35 – 500	209
Carbohydrate	252 – 931	522
P	11 – 160	50
Cl	48 – 469	276
Ca	57 – 112	37
Mg	25 – 49	87
N	60 – 807	322
K	11 – 160	67

Source: Second National Symposium on Food Processing Wastes (1971)

To interpret the data of these and other refractory elements in dairy wastes,
values for calcium, potassium, sodium, phosphorus, magnesium, chloride
and nitrogen in milk wastes containing 0.1 percent (1,000 ppm BOD_5) milk
solids would be as shown on the following page.

Material	PPM
BOD	1,000
Nitrogen	55
Chloride	10
Phosphorus	12
Calcium	12
Sodium	4
Potassium	15
Magnesium	5

Available data for magnesium, calcium, sodium and potassium are also summarized in Table 25. The values differ widely in respect to sodium (60 to 807 ppm), and potassium (11 to 160 ppm) and were less variable for magnesium (25 to 49 ppm), and calcium (57 to 112 ppm). Variations in the latter two minerals reflected variations in BOD, whereas variance in sodium and potassium were unrelated to BOD and reflected differences in the sodium and potassium composition of cleaning compounds and water treatment utilized in the different plants.

Based on these data, milk salts would contribute about 10 to 50%, 10 to 75%, 1 to 10%, and 25 to 100% of the magnesium, calcium, sodium, and potassium ions, respectively, in dairy food plant wastes. Thus, 90% of the sodium, 50% of the magnesium and 25% of the calcium appeared to come from nondairy sources, presumably cleaning compounds, detergents and lubricants.

Lubricants are not mentioned in the literature as a source of pollutant in dairy plant wastewaters, only limited data could be obtained during the course of this study concerning specific contribution of these materials to BOD. Most of the lubricants are salts of fatty acids and range in BOD from 0.8 to 1.2 pounds of BOD_5 per pound of material.

Detergents and Sanitizers

A significant contribution to the water pollution problems presented by the dairy products industry is made by detergents and sanitizers employed in plant cleaning operations.

Selected compounds used in the manufacture of detergents and related materials and their contribution to BOD_5 (lb./lb. of product) are shown in Table 26. Wetting agents and surfactants varied widely in BOD values ranging from 0.05 to 1.2. The most commonly utilized surfactant contained 0.65 pound of BOD_5/lb. of product; nonionic wetting agents exhibit a low BOD, with average values of 0.2 pound BOD per pound; and acids which are used in dairy food plant detergents, have BOD values ranging from 0.25 to 0.85. The most commonly used acid detergent in cleaning food plant equipment has a BOD value of 0.65 pound BOD_5 per pound of detergent.

TABLE 26: BOD_5 OF SELECTED CHEMICALS IN DETERGENTS, SANITIZERS, AND LUBRICANTS USED IN DAIRY FOOD PLANTS

Material	Lb. BOD per lb. Product
Acetic acid	0.65
Duponol D, alkyl alcohol, sulfonated	0.45
70% hydroxyacetic acid	0.07
Alkyl phenyl condensate of ethylene oxide	0.04
Phenoxypolyoxyethylene	0.005
Nacconol NR-Na alkylarylsulfonate	0.004
Neutronyx 600, aromatic polyglycol ether	0.0
Nopco 1111-sulfonated coconut oil	0.96
Nopco 1665-soluble fatty acid ester	0.12
Pine oil	1.08
Tallow	1.52
Triethanolamine	0.01
Ultrawet DS-sodium alkylarylsulfonate	0.0
Linear alkylarylsulfonate	0.65
Ethylene glycol	0.70
Zalon-fatty amide	0.20

Source: Second National Symposium on Food Processing Wastes (1971)

The amount of various detergent components used in cleaning various types of dairy food plant by hand and CIP method were calculated and are listed in Table 27 in lb./1,000 lbs. of milk processed. Alkali, as sodium hydroxide, is the major component ranging from 0.23 to 1.52 lbs./1,000 lbs. of milk processed for hand cleaning and from 0.23 to 0.47 for CIP. Phosphate values ranged from 0.05 to 0.76 and 0.04 to 0.21 for CIP and hand cleaning, respectively. Acid ranged from 0.08 to 0.84 for hand cleaning and from 0.07 to 0.15 for CIP, whereas surfactant was the component used in the lowest concentrations, with amounts per thousand pounds of milk processed ranging from 0.02 to 0.11 for both hand and CIP cleaning, respectively.

Ice cream plants utilized the highest concentration of detergents, with the least differences between hand and CIP cleaning procedures. This reflects the relative amount of equipment in the ice cream plants still cleaned by hand and the fact that CIP is limited to pipelines and storage tanks.

TABLE 27:   COEFFICIENTS OF COMPONENTS OF CLEANERS AND SANITIZERS, LB. PER 1,000 LBS. MILK PROCESSED

Product	Market Milk		Butter		Cheese		Condensed & Powder		Ice Cream	
	hand	CIP	hand	CIP	hand	CIP	hand	CIP	hand	CIP
Alkali[*]	0.77	0.25	0.79	0.25	0.65	0.41	0.23	0.21	1.52	0.47
Phosphate**	0.12	0.05	0.76	0.15	0.38	0.07	0.046	0.04	0.22	0.22
Surfactant	0.08	0.02	0.025	0.08	0.08	0.05	0.02	0.02	0.11	0.11
Acid	0.84	0.15	0.35	0.10	0.14	0.09	0.088	0.07	0.52	0.15
Chlorine	0.34	0.10	0.14	0.08	0.08	0.03	-----	----	0.37	0.13

*NaOH
**Hexa-meta PO_4

Source: Second National Symposium on Food Processing Wastes (1971)

Only acids and surfactants contribute BOD to the wastewater. The amount of BOD_5 in the wastewater in lb./1,000 lbs. processed from materials used in dairy food plant detergents as a function of concentration is shown in Figure 10. Under average conditions in the modern milk plant, the amount of BOD contributed by surfactant and acid detergent could be about 0.1 pound/1,000 pounds of milk processed. The organic acid appears to be the major source of BOD and could be substituted by an inorganic acid. However, the major inorganic acid of choice will be phosphoric acid which would add substantially to phosphate levels.

Data for cleaning and sanitizing solution in dairy plant wastewater for 27 plants are summarized in Table 28. Cleaning solution in percent of wastewater, ranged from 2.2 to 41.6% with an average value of 15%. The sanitizer solution contributed 0.2 to 13.8% of the wastewater and averaged 3.1%. The lower the wastewater coefficient for a dairy plant, the greater was the percentage of that wastewater made up of detergent and sanitizer solutions.

The average concentration of the detergent in the cleaning solution would be 5,000 ppm and for sanitizers the average would be 100 ppm. In terms of the wastewater, the concentration of alkali, sodium hydroxide, ranged from 32 to 6,280 ppm with an average of 500 ppm; phosphate ranged from 30 to 75 ppm with an average of 43 ppm; acid ranged from 10 to 348 ppm and averaged 66 ppm; whereas chlorine ranged from 6 to 234 ppm and averaged 70 ppm.

FIGURE 10: CONTRIBUTION OF DETERGENT MATERIALS TO BOD

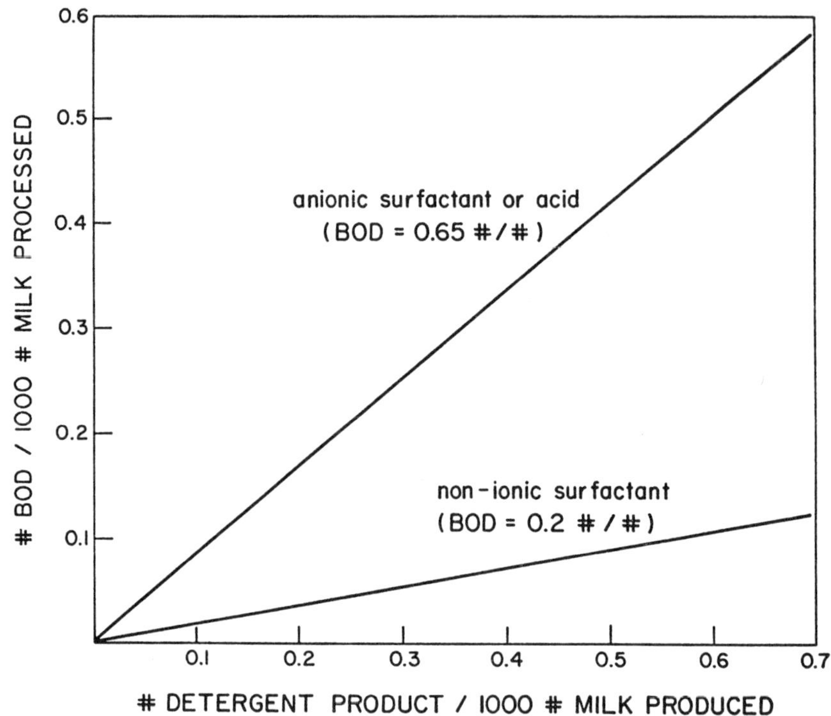

Source: Second National Symposium on Food Processing Wastes (1971)

Based on this data, it appears that detergents and sanitizers may be more significant in dairy wastes than previously reported (3). They potentially contribute significantly to BOD, to refractory COD and may be significant in providing toxicity and poor performance to dairy waste treatment facilities. Their effect on the dominating microflora of dairy food plant waste treatment systems and a full determination of their real significance requires further investigation.

The volume and concentration of sanitizers utilized in the sanitation of dairy plant equipment by nonrecycle systems are shown in Table 29. Since cleaners and sanitizers appear to have an effect on the refractory material in dairy plant wastes and contribute materially to COD and, to a lesser extent, BOD, information is needed in respect to the volume and concentration of materials utilized. A comparison between the concentration and volume for recycle and nonrecycle treated systems for two fluid milk and cottage cheese plants of the same size is shown in Table 30. The systems are essentially equal in volume used, whereas the recycle system used 35% more cleaning compound to achieve equivalent cleaning efficiency.

TABLE 28:   CLEANING AND SANITIZING COMPOUNDS

| Type of Plants | % Waste Water as | | | | Average ppm of Following | | | | |
| | Cleaning Solut. | | Sanitizing Solut. | | | | Phos- | Sur- | |
	Range	Aver.	Range	Aver.	Alkali	Acid	phate	factant	Cl
Milk	3.4–21	9.4	0.6–5.0	2.2	353	51	175	52	155
Cheese	4.6–22	14.3	5.0–6.2	5.8	---	--	---	--	52
Ice Cream	2.22	12.0	0.2–3.0	1.3	43	33	10	5	16.8
Cottage Cheese	---	18.0	---	10	150	75	64	21	234
Milk & Cottage Cheese	7–41	17.0	0.3–13.8	3.4	360	55	56	41	102
Milk, Cottage Cheese and Ice Cream	10–22	15.6	1.4–5.0	3.0	184	34	73	25	57

Source: Second National Symposium on Food Processing Wastes (1971)

TABLE 29:   SANITIZER UTILIZATION OF DAIRY FOOD PLANT EQUIPMENT

Process	Pounds Sanitizing Solution/1000 Pounds Milk Processed	ppm Sanitizer Required
Tank trucks	0.16	50
Storage tanks (20,000 gal.) lines	0.006	50
HTST Pasteurizer	0.0005	50
Fillers	0.001	100

Source: EPA Report 12060 EGU, March 1971

TABLE 30:   COMPARISON OF CLEANING SYSTEMS IN TWO FLUID
MILK-COTTAGE CHEESE PLANTS OF THE SAME SIZE*

Plant	Volume Milk Processed	Pounds of Cleaning Compounds Used/Day	Gallons of Cleaning Solution/Day	Average Percent Concentration of Cleaning Solution
A	530,000 lb	699	18,000	0.4
B	500,000 lb	515	21,000	0.2

*Plant A used a recycle cleaning system, recharging six CIP or COP tanks
twice a day to accomplish efficient cleaning; Plant B Used the "once-
through" or nonrecycle system of cleaning with two centralized CIP
tanks and two COP tanks.

Source: EPA Report 12060 EGU, March 1971

PROCESS SOURCES OF
DAIRY FOOD PLANT WASTES

Determination of the significant sources of dairy food plant wastes requires an understanding of the processing of dairy foods, the various unit operations and their potential role as sources of wastewater, milk solids and refractory compounds. Flow diagrams of various unit processes for processing milk into its various products, waste generation processes and the general nature of the waste for each type of unit operation is represented in flow diagrams which accompany the text which follows.

The waste generating processes of major significance in the dairy food industry include:

      washing, cleaning and sanitizing of all pipelines, pumps, processing equipment, tanks, tank trucks and filling machines,

      startup, product changeover and shutdown of HTST and UHT pasteurizers,

      loss in filling operations through equipment jams and broken packages,

      lubrication of casers, stackers and conveyors.

In the flow diagrams, the piping associated with a unit process is considered to be an integral part of that operation. Thus, the waste generating process of tank truck washing would include all pumps and piping in the receiving room; cleaning of raw storage tanks would include the associated raw milk pumps, valves and lines.

Spills and leaks from pipeline joints, valves and pumps have not been included as waste generating processes in any of the flow diagrams. Although these sources are significant to varying degrees in different dairy food plants, they are inherently controllable through the application of good management practices. Generally, the nature of the wastes generated by the various processes are similar; including milk solids in some form, plus detergents and sanitizers in most product processing steps and these same materials plus lubricants are generated in packaging and distribution operations. The significance of various waste generating processes to the total waste load of the dairy food plant wastewater is a characteristic of the individual process.

Literature values, for BOD_5 from various unit operations which are in general agreement with each other, have been presented by two investigations (19) (20). Values obtained from several modern, automated dairy food plants manufacturing fluid milk product and cottage cheese are in Table 31.

Exclusive of whey and washwater, the process yielding the highest BOD_5 per 1,000 lbs. of milk processed is pasteurization; with the milk solids coming from the startup, changeover and shutdown of the process on water. The next most significant source is the filling operation, where machine jams in high speed packaging equipment are frequent. The loss of 5 to 10 gallons of milk in a single machine jam is not uncommon.

TABLE 31:   INDUSTRIAL DATA FOR BOD COEFFICIENTS FOR UNIT PROCESSES IN TWO FLUID MILK COTTAGE CHEESE PLANTS

Process	Lbs. of BOD/1,000 Lbs. of Milk Received	
	Plant A	Plant B
Tank truck rinsings	0.25	
Silo tank rinsings	0.23	
Tank trucks and storage tank rinsings		0.25
Separation (CIP)	0.15	
CIP Separator sludge		0.08
HTST Startup and changeover and cleaning	0.75	0.55*
Filling operations – milk	0.30	0.32
Lubricants		0.08
Product returns		0.10
Cottage cheese washwater	0.33	0.60
Unaccounted	0.20	0.03
Total wastewater	2.21	2.01

*Initial startup collected, product changeovers not diverted.

Source: Second National Symposium on Food Processing Wastes (1971)

Limited data for ice cream operations indicated that the loss of BOD_5 in rinsing mix tanks and ice cream freezers ranged from 2 to 10 lbs. of BOD_5 per 1,000 lbs. of milk processed into ice cream. Because of the high viscosity of the mix and ice cream, the percentage of product remaining on the surface after draining can be up to 8.5% of the total present in the processing equipment.

Based on industry observation and evaluation of literature data, volume and BOD coefficients per 1,000 lbs. of milk processed (or milk equivalent) for various operations are presented in a second series of flow diagrams which accompany the text which follows. The data should be considered as guides only, which would require a relatively good management to achieve, and may be expected to vary from plant to plant. Much more data on specific unit operations of different type plants measured under controlled conditions is needed to provide a truly reliable index of the role of various unit operations in waste loading.

In determining the over-all waste coefficients for various types of plants, corrections were made to compensate for the amounts of wastewater or BOD generated by the various processes for each unit weight of milk processed. For example, in cheese operations, the starter manufacturing process generates about 0.1 pound of water per pound of product and about 0.3 pound of BOD per 10,000 pounds of milk processed. However, starter is used at the rate of 1.0%; thus, this operation would contribute only 0.01 pound of water per pound of milk processed and 0.03 pound of BOD per 1,000 pounds of milk processed into cheese.

The coefficients for all operations can be reduced in the same manner as will be illustrated for fluid milk operations. The values reported by some of the plants surveyed during the study reported by Harper et al (2) had coefficients materially below those reported in the various flow sheets.

Interim effluent guidelines for various industries were issued by the Office of Permit Programs of the Environmental Protection Agency on May 1, 1973. These effluent limitations are shown in Table 32. The following assumptions regarding processing were made in the development of the guidelines:

> good degree of management effort in controlling available product and raw material losses,
>
> recovery of whey and by-products,
>
> biological oxidation with trickling filter or activated sludge,
>
> secondary clarification,
>
> disinfection, if necessary.

The system described on the preceding page is a generalized model which is applicable to the entire industry. At the present time there are very few alternatives to this model. Of major importance in this model is the control of the production process.

TABLE 32:   EFFLUENT LIMITATIONS IN POUNDS PER 1,000 POUNDS MILK OR MILK EQUIVALENT

	Schedule A		Schedule B	
Subcategory	BOD_5	Suspended Solids	BOD_5	Suspended Solids
Receiving Stations	0.032	0.032	0.053	0.053
Fluid Milk Processing	0.03	0.03	0.13	0.13
Butter	0.03	0.03	0.13	0.13
Natural Cheese	0.03	0.03	0.20	0.20
Ice Cream*	0.03	0.03	0.45	0.45
Condensed Milk	0.03	0.03	0.11	0.11
Dry Milk	0.03	0.03	0.11	0.11
Whey Condensing	0.03	0.03	0.15	0.15
Whey Drying	0.03	0.03	0.20	0.20
Cottage Cheese	0.13	0.13	0.95	0.95

NOTE: *Novelty items (e.g., stick confections, popsicles) shall be considered on a case-by-case basis.

Source: Environmental Protection Agency

As a production basis for calculation of allowable effluent limitations, the maximum daily receipts in 1,000 pounds milk equivalent, shall be the highest average level sustained for seven consecutive operating days of normal production, determined at the time of application. (Milk equivalent may be determined from Table 33.)

TABLE 33:   DETERMINATION OF MILK EQUIVALENT

Milk equivalent is the quantity (in pounds) of milk used to produce one pound of product. Milk is considered to include skim milk, whole milk, and other milk with a butterfat content of less than 3.5%. If the facility receives higher than 3.5% butterfat milk or cream as a raw product, an adjustment according to the butterfat ratio should be made. In the case of cream, this ratio would be 48/3.5; 1,000 lbs. of cream would therefore be equivalent to 13,700 lbs. of milk.

(continued)

TABLE 33:   (continued)

If the raw product received is whey, the milk equivalent should be considered as the BOD ratio of whey to whole milk, (i.e., 0.4); therefore, 1,000 lbs. of whey is equivalent to 400 lbs. of milk. If the pounds of final product are available but no raw material data are known, the following data can be used to convert the final product to its milk equivalent.

Product	Lbs. Milk Equivalent per Lb. Product
Butter	21.3
Natural cheese	9.9
Cottage cheese	6.3
Ice cream	2.7
Condensed milk	2.4
Dry whole milk	7.4
Dry skim milk	11.0

Source: Environmental Protection Agency

FLUID MILK OPERATIONS

Flow diagrams 11 through 15 are for fluid milk operations, their subprocesses and various alternative processes for certain by-products. Milk receiving and storage are common to all of the unit operations and their subprocesses. Therefore, waste volume for BOD coefficients prepared for receiving and storage operations apply equally to all fluid milk processes and subprocesses and apply also to other dairy food plants receiving milk by tank truck. Other unit operations that apply to all dairy food plants include the utilization of centrifugal machines, (for clarifying, separating and/or standardizing), pasteurized product, and filling and distribution.

In market milk processing, the significant waste generating processes include tank washing, sludge centrifugal machines, tank washing and sanitizing, high-temperature short-time startup and product changeover cleaning, high-temperature short-time period cleaning and sanitizing of processing vats and lines, cleaning and sanitizing of storage vats, drips, broken packages, damaged packages, cleaning of fillers, conveyor lubrication and cleaning, broken packages, conveyor lubrication and returns constitute waste generated steps in cold storage and distribution.

FIGURE 11: MARKET MILK FLOW DIAGRAM

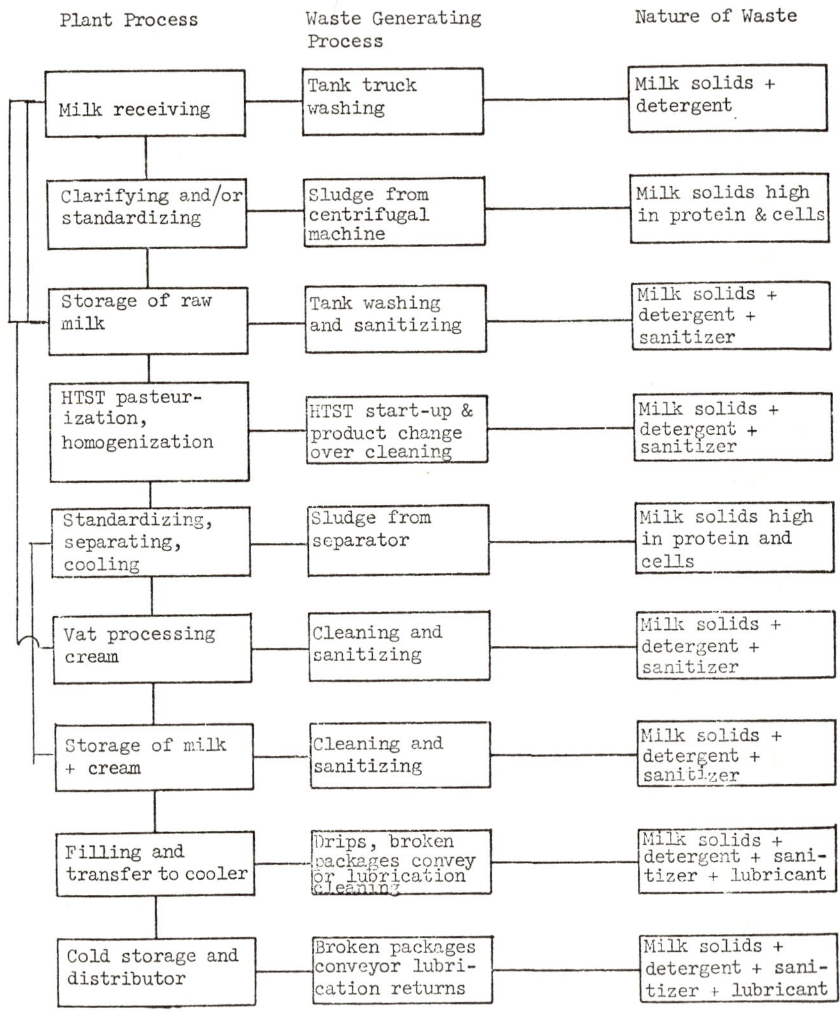

Source: EPA Report 12060 EGU, March 1971

FIGURE 12: FLOW DIAGRAM FOR MARKET MILK SUBPROCESS — SKIM MILK, CREAMS AND SPECIAL MILKS BY BATCH PROCESSING (ALTERNATE 1)

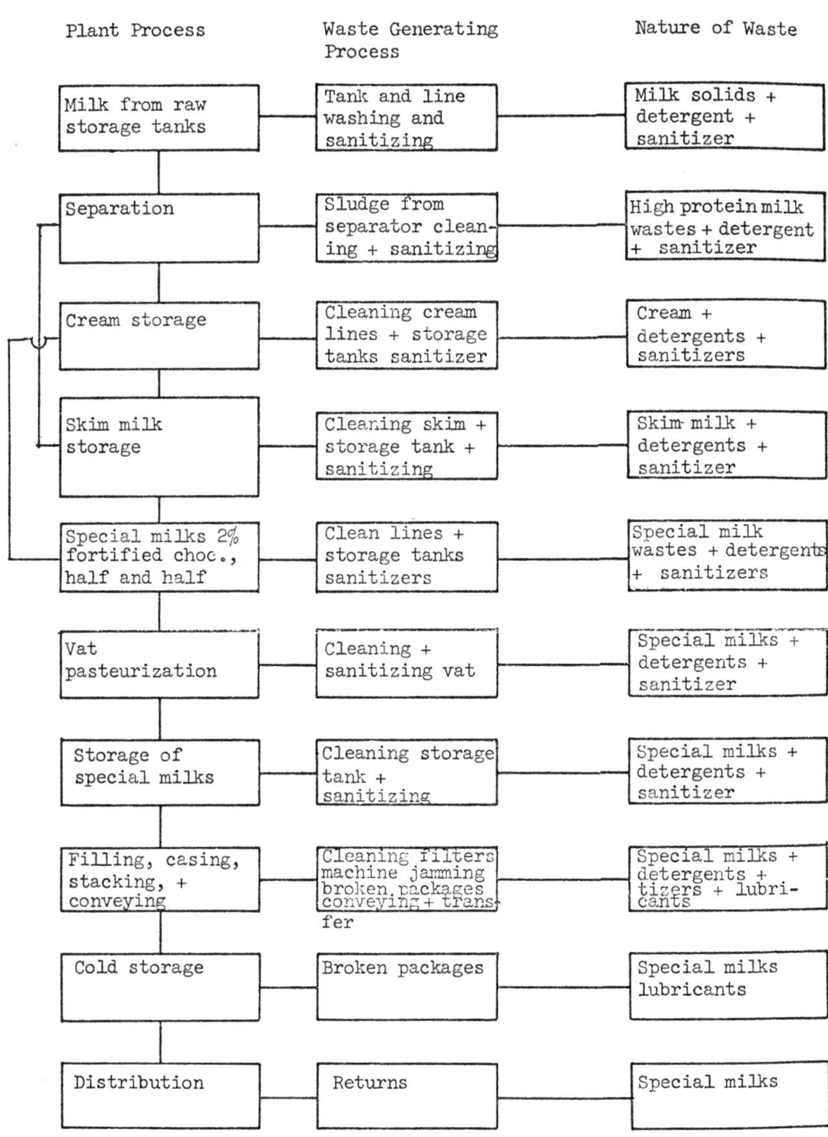

Source: EPA Report 12060 EGU, March 1971

FIGURE 13:   FLOW DIAGRAM FOR MARKET MILK SUBPROCESS —
SKIM MILK, CREAMS AND SPECIAL MILKS BY CONTINUOUS PROCESS
(ALTERNATE 2)

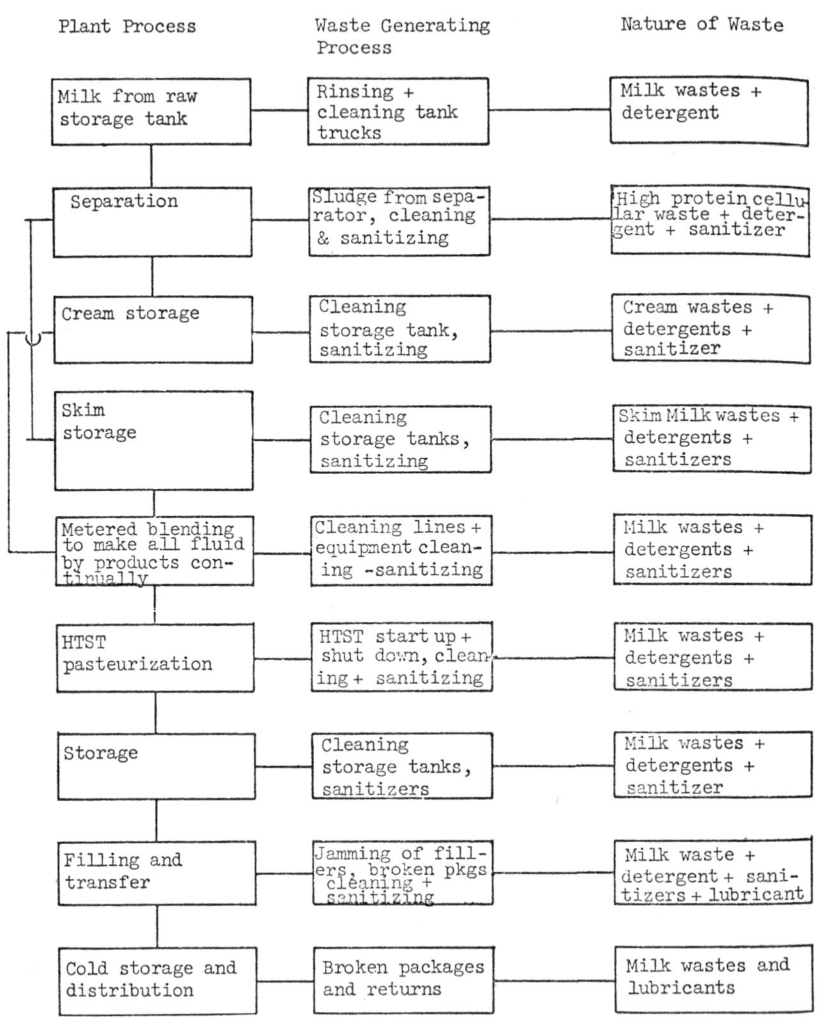

Source: EPA Report 12060 EGU, March 1971

FIGURE 14: FLOW DIAGRAM FOR MARKET MILK SUBPROCESS — SKIM MILK, CREAMS AND SPECIAL MILKS BY CONTINUOUS PROCESS (ALTERNATE 3)

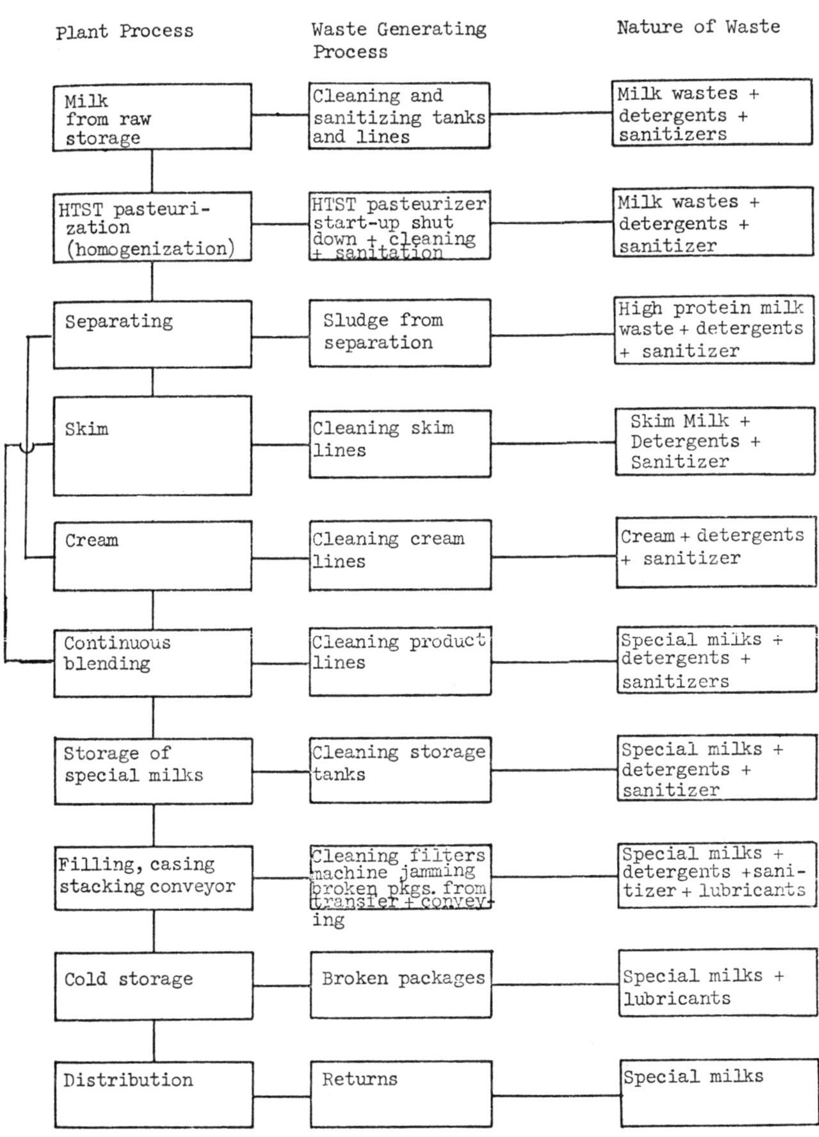

Source: EPA Report 12060 EGU, March 1971

FIGURE 15:   FLOW DIAGRAM FOR MARKET MILK SUBPROCESS — BUTTERMILK, YOGURT AND SOUR CREAM

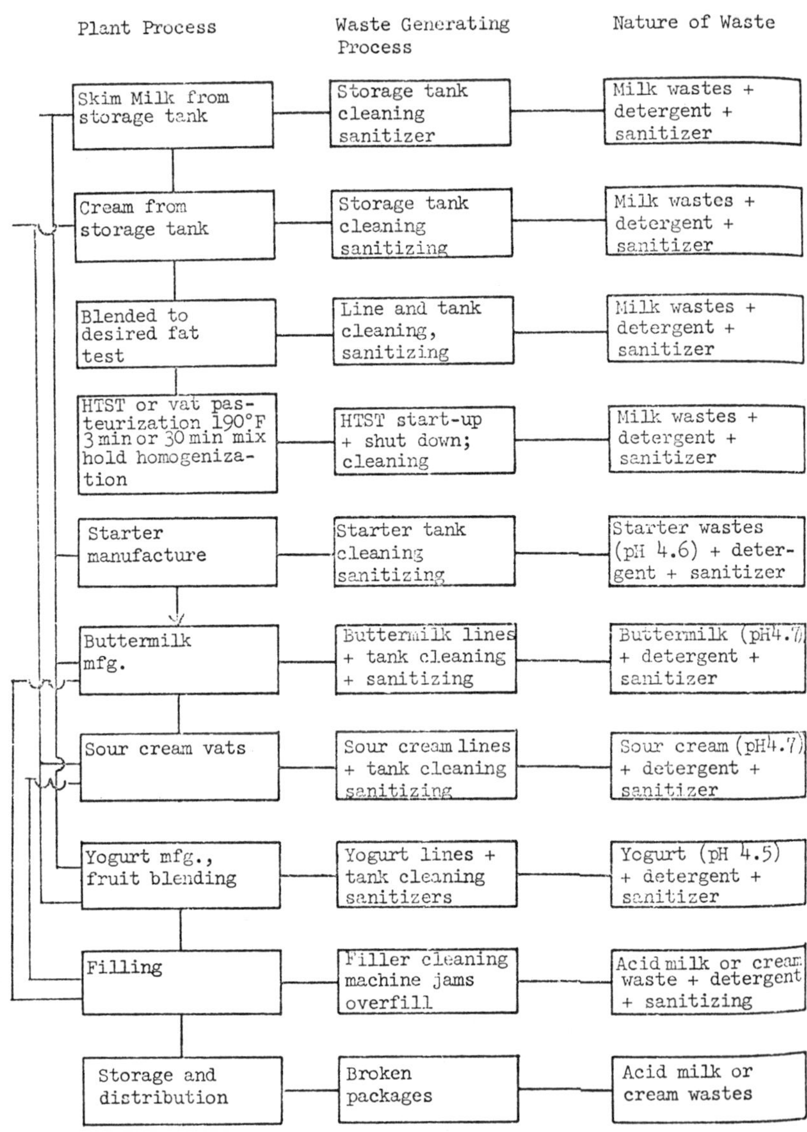

Source: EPA Report 12060 EGU, March 1971

Another type of portrayal of the wastes generated from fluid milk processing is given in Table 34. The key to the abbreviations used are given in Table 35. An itemization of the wastes from buttermilk, yogurt and sour cream manufacture is given in Table 36.

TABLE 34: TYPE OF WASTES FROM FLUID MILK PRODUCTS PROCESSING

Plant Operation	Waste Generation From	Type of Waste
Receiving	Wash and sanitize raw milk lines	MS, DET, SAN
Clarifying	Sludge from clarifier	MS (High Protein)
RM Storage	Wash and sanitize tanks	MS, DET, SAN
Standardizing, Blending mixing, storage	Wash and sanitize tanks and lines, sludge from separator	MS, DET, SAN
HTST	Start-up, product change over, cleaning	MS, DET, SAN
Transfer and storage	Wash and sanitize tanks and lines	MS, DET, SAN
Filling	Drips, broken packages, cleaning, conveyors	MS, DET, SAN, LUB
Transfer to cooler	Broken packages conveyors	MS, LUB
Distribution	Broken packages conveyors returns	MS, LUB

Source: Report PB 220,704

TABLE 35: GLOSSARY OF ABBREVIATIONS

C	Cream	LUB	Lubricants
CC	Cottage cheese	M	Milk
CIP	Cleaning in place	MS	Milk solids
DET	Detergents	RM	Raw milk
HTST	High-temperature short-time system (pasteurization and homogenization)	S	Skim
		SAN	Sanitizers
ICM	Ice cream mix		

Source: Report PB 220,704

TABLE 36:   TYPE OF WASTES FROM BUTTERMILK, YOGURT AND SOUR CREAM

Plant Operation	Waste Generation From	Type of Waste
SM, C	Wash and sanitize lines and tanks	MS, DET, SAN
Standardize, blend	Wash and sanitize lines, tank and separator, sludge from separator	MS, DET, SAN
Pasteurize	Wash and sanitize HTST or vat	MS, DET, SAN
Starter Manufacture	Clean lines and tank	Starter (pH 4.6), DET, SAN
Hold		
For Yougurt – Blend fruits and flavors	Wash blend tank	Fruits, flavors, SAN, DET
Filling	Clean Filler, Broken packages, jammed machine	DET, SAN, Product Wastes (pH 4.5–4.7)

Source: Report PB 220,704

Figure 16 depicts the wastewater volume and BOD_5 coefficients for the various unit operations of the fluid milk operation under good management, with no extra steps taken to eliminate wastes.

Figure 17 shows the same process after the elimination of all rinses, product from startup, changeover and shutdown of the HTST unit, collection of all spillage from filling operations and elimination of all returns from the waste stream. The resulting coefficients are < 0.5, and indicate what can be done through management and engineering control.

Figures 18 through 21 show wastewater volume and BOD_5 coefficients for the processes shown in the block flow diagrams in Figures 12 through 15.

FIGURE 16:   WASTE COEFFICIENTS FOR MARKET MILK — NORMAL OPERATION

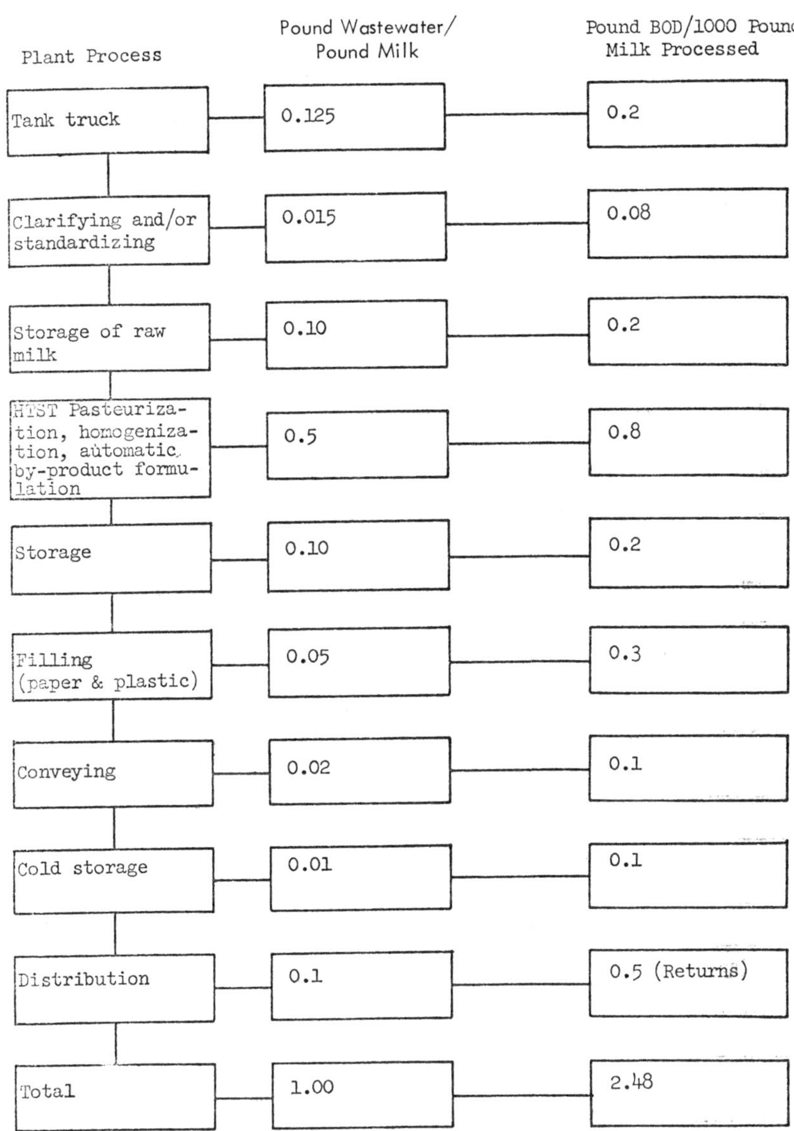

Source: EPA Report 12060 EGU, March 1971

FIGURE 17:   WASTE COEFFICIENTS FOR MARKET MILK PROCESSING *

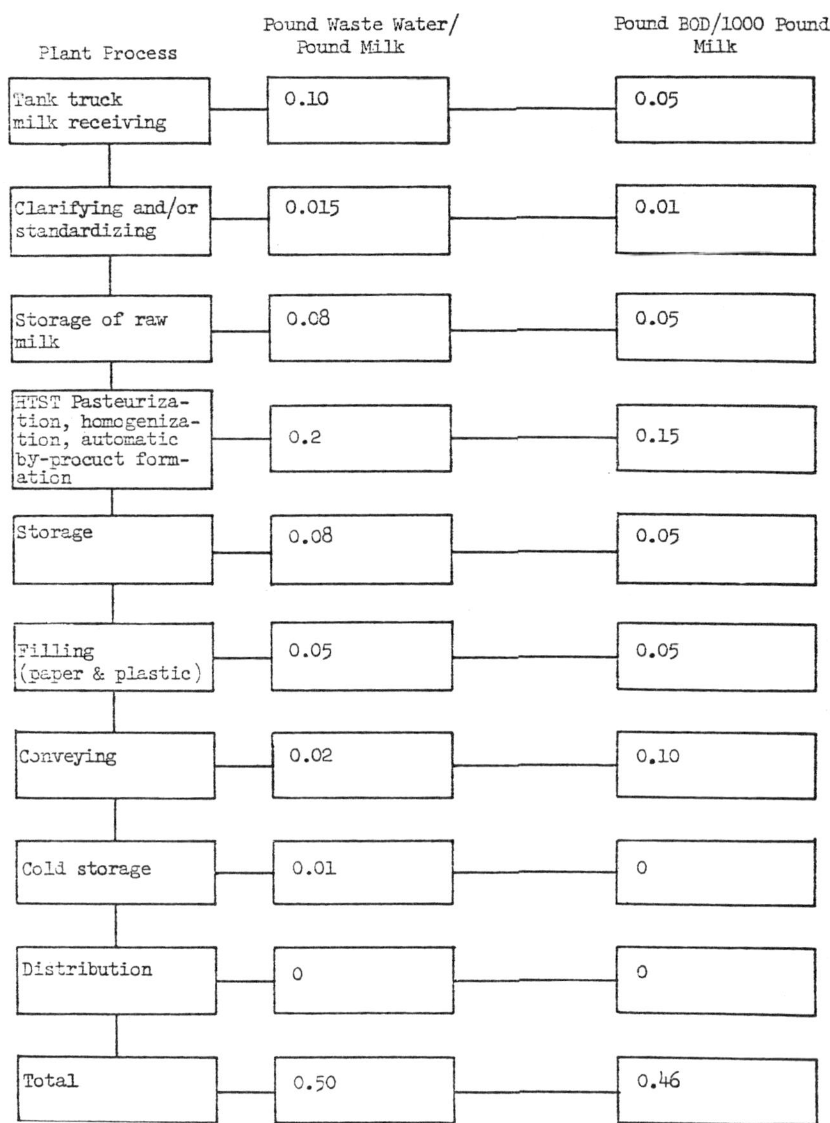

*Rinses saved, CIP sludge saved, HTST startup, changeover and shutdown segregated and saved; returns used as feed.

Source: EPA Report 12060 EGU, March 1971

FIGURE 18: WASTE COEFFICIENTS FOR MARKET MILK SUBPROCESS — SKIM MILK, CREAMS AND SPECIAL MILKS BY BATCH PROCESSING (ALTERNATE 1)

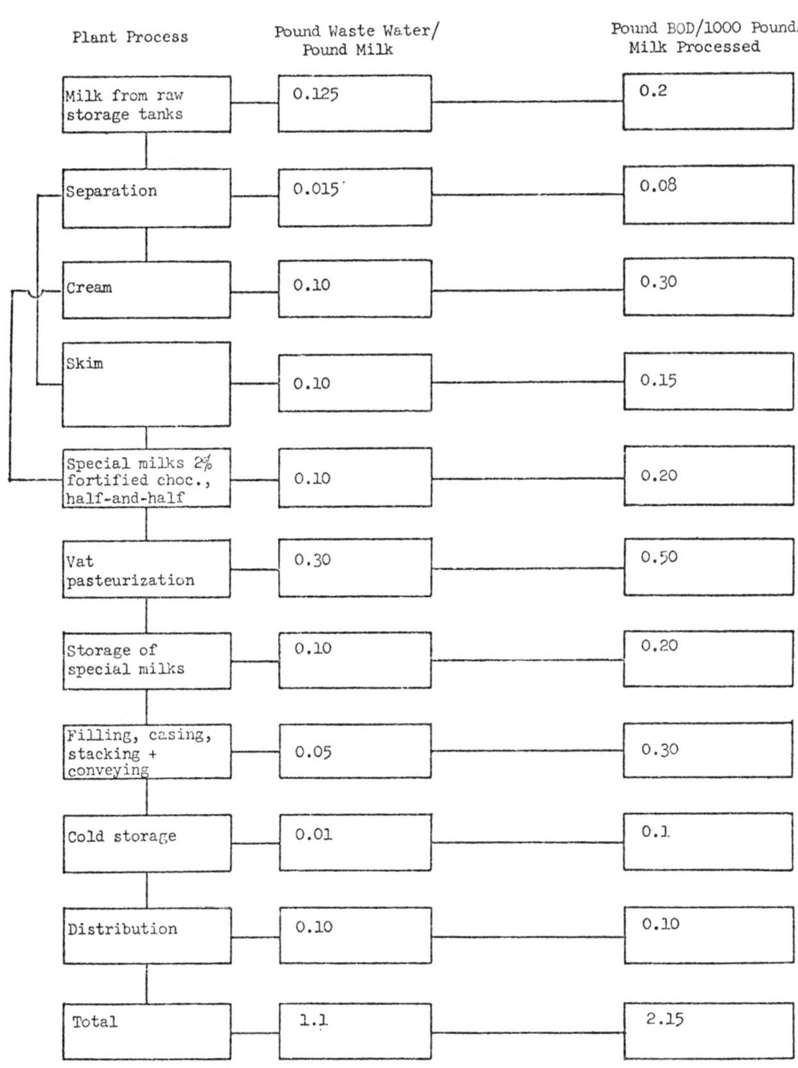

Source: EPA Report 12060 EGU, March 1971

FIGURE 19:   WASTE COEFFICIENTS FOR MARKET MILK SUBPROCESS —
SKIM MILK, CREAMS AND SPECIAL MILKS BY CONTINUOUS PROCESS
(ALTERNATE 2)

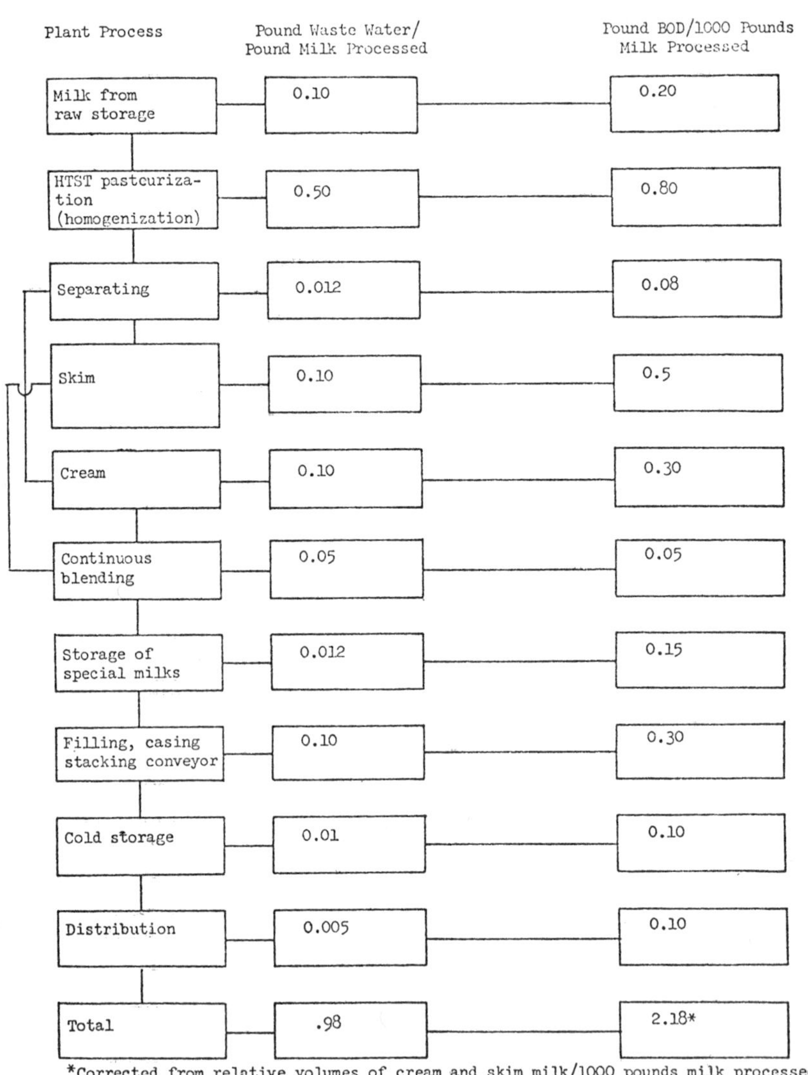

Source: EPA Report 12060 EGU, March 1971

FIGURE 20: WASTE COEFFICIENTS FOR MARKET MILK SUBPROCESS — SKIM MILK, CREAMS AND SPECIAL MILKS BY CONTINUOUS PROCESS (ALTERNATE 3)

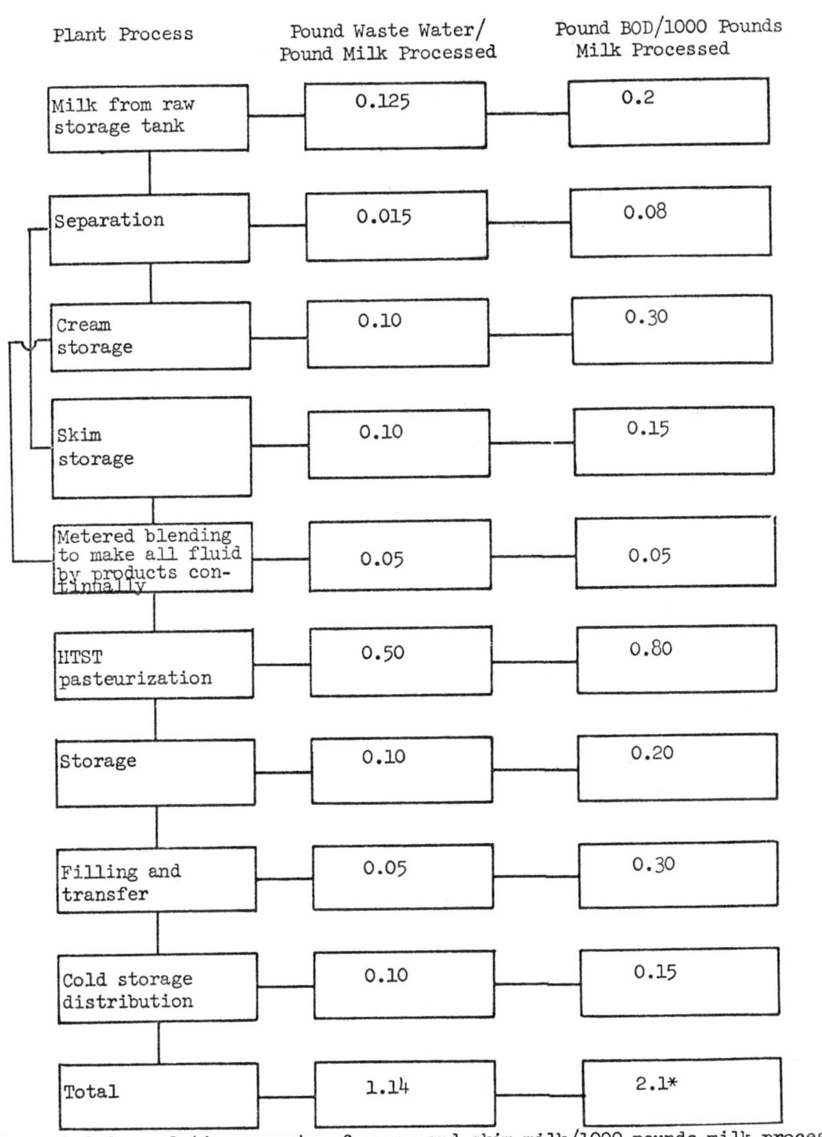

*Corrected for relative amounts of cream and skim milk/1000 pounds milk processed

Source: EPA Report 12060 EGU, March 1971

FIGURE 21:   WASTE COEFFICIENTS FOR MARKET MILK SUBPROCESS —
BUTTERMILK, YOGURT AND SOUR CREAM

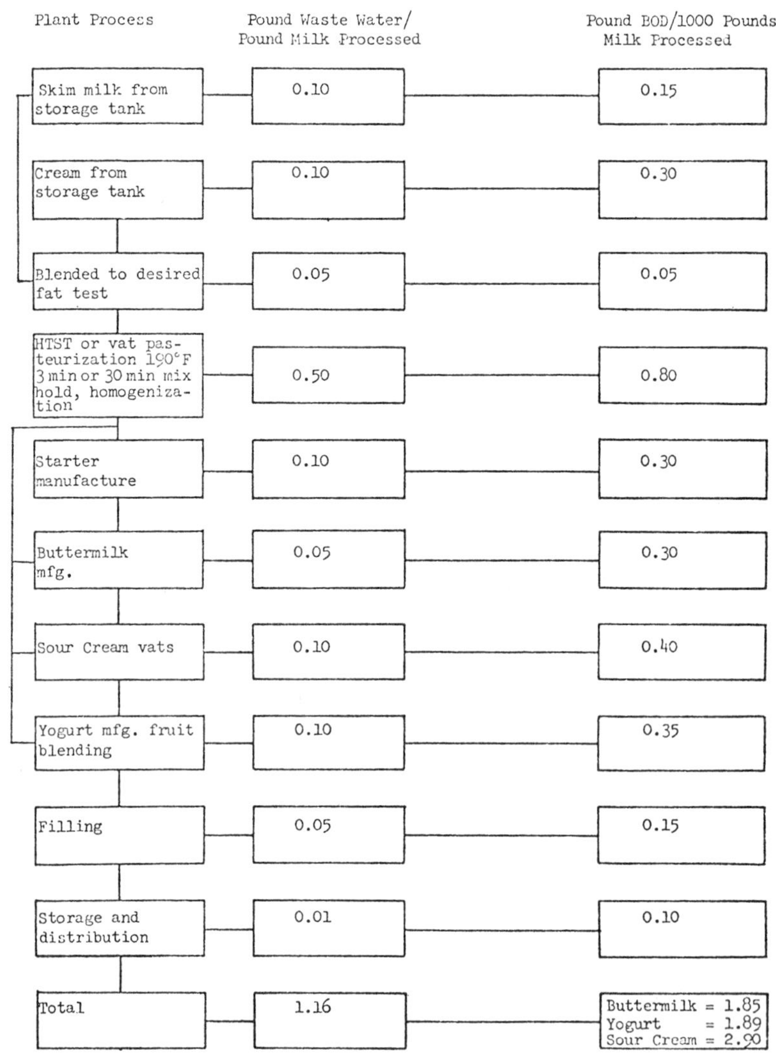

Source: EPA Report 12060 EGU, March 1971

BUTTER

Figures 22 and 24 give the outlines of the churn and the continuous method
for butter operation. The continuous process materially reduces potential
waste strength by eliminating the buttermilk production and the washing
steps. Disposition of skim milk and buttermilk for food use may be accomplished by the use of a drying operation in conjunction with the butter
operation, or these materials may be shipped to another dairy food plant
by tank truck.

FIGURE 22:   FLOW DIAGRAM OF BUTTER BY CONTINUOUS PROCESS

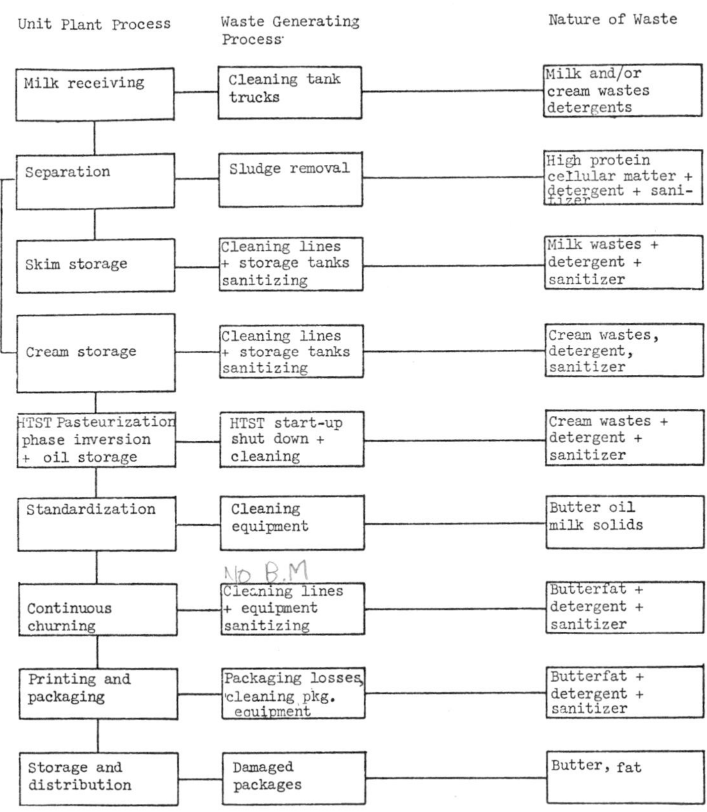

Source: EPA Report 12060 EGU, March 1971

Figures 23 and 25 show wastewater volumes and BOD$_5$ coefficients for the corresponding processes.

FIGURE 23:   WASTE COEFFICIENTS FOR BUTTER BY CONTINUOUS PROCESS

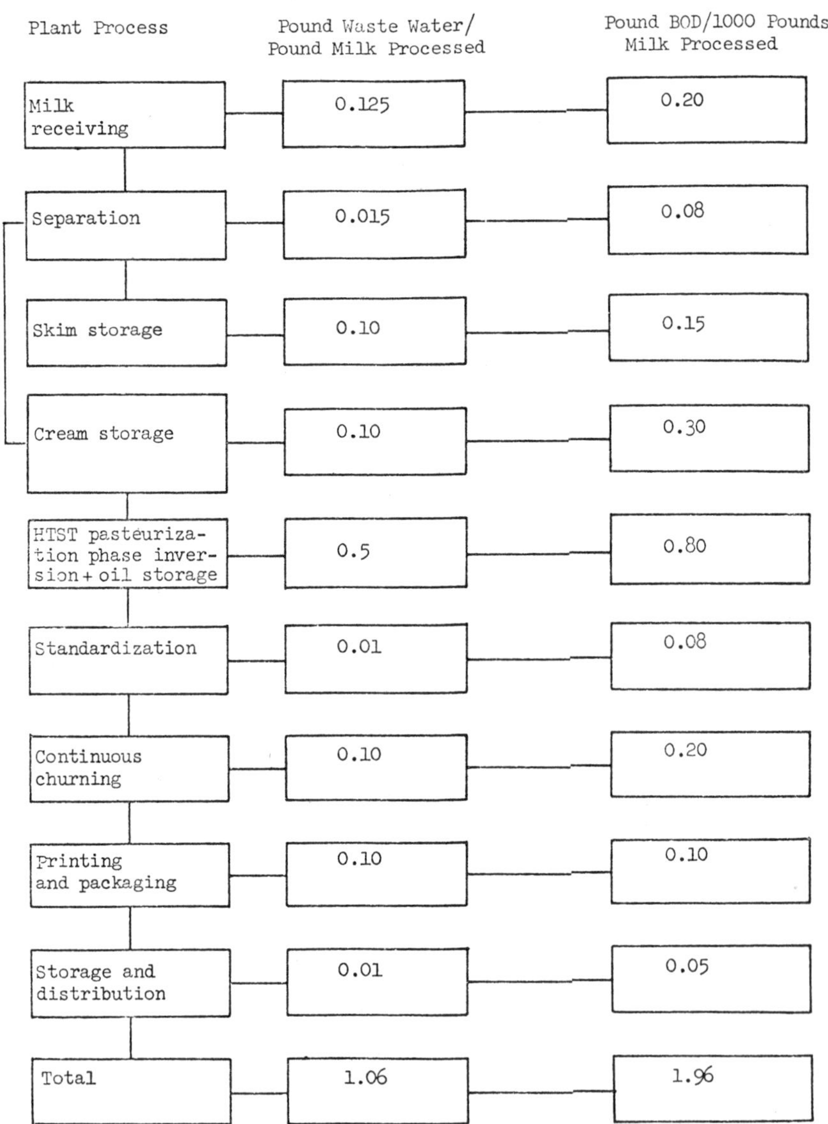

Source: EPA Report 12060 EGU, March 1971

FIGURE 24: FLOW DIAGRAM OF BUTTER BY CHURN PROCESS

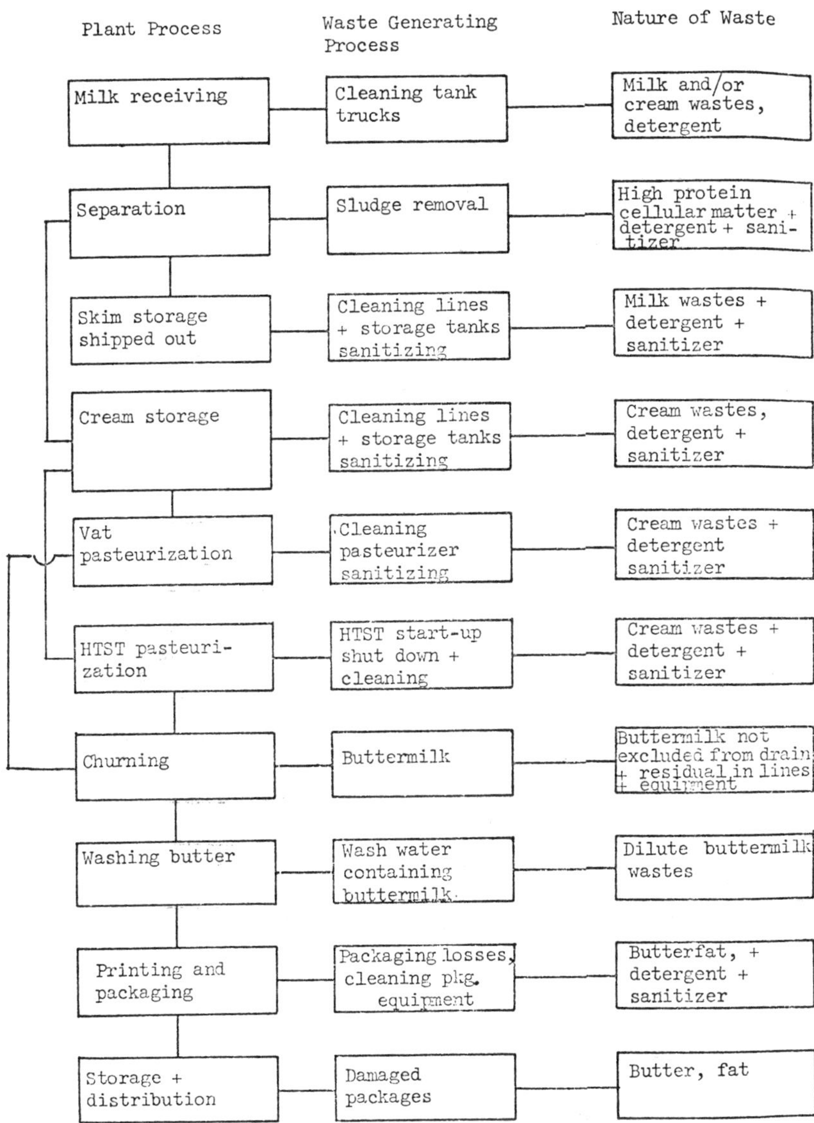

Source: EPA Report 12060 EGU, March 1971

FIGURE 25: WASTE COEFFICIENTS FOR BUTTER BY CHURN PROCESS

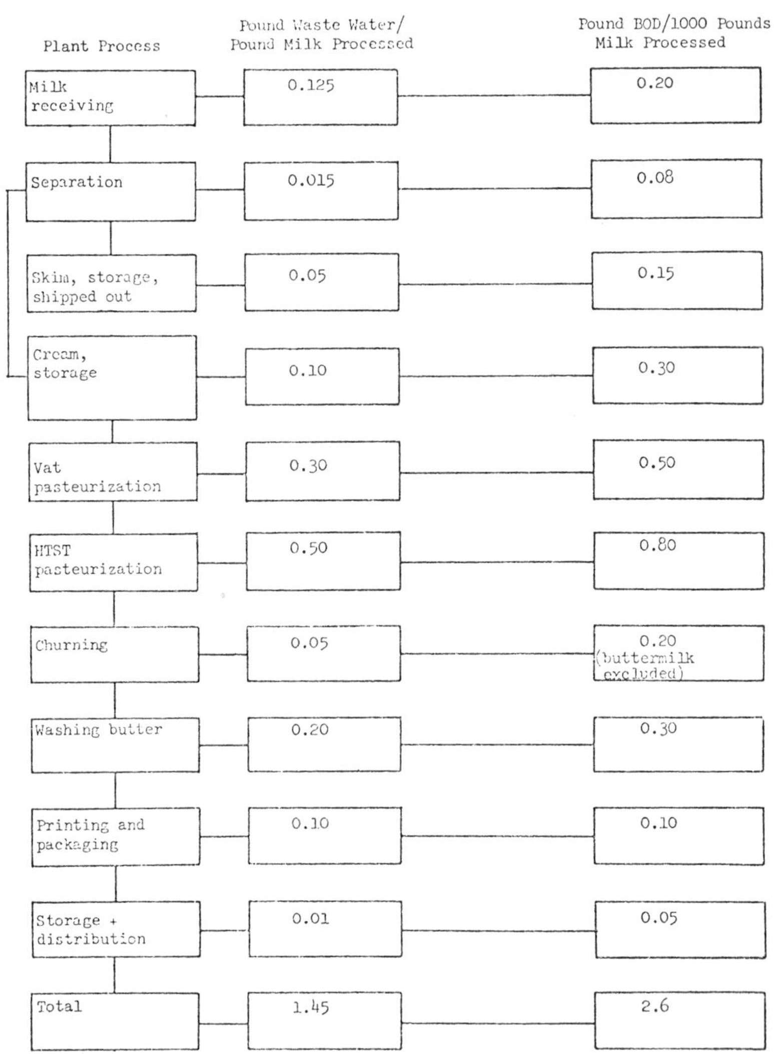

Source: EPA Report 12060 EGU, March 1971

CHEESE

Unit process operations, waste generating processes and the nature of wastes for cottage cheese, cheddar cheese, Swiss cheese and Italian cheese operations are outlined in Figures 27, 29, 31 and 32 to represent different types of cheese manufacturing processes.

The most visible source of waste in the dairy industry is in the whey resulting from the various cheese operations. There has been an increasing trend in the industry to exclude all whey from waste treatment and to use it for feed or food. Where whey is to be shipped to another plant for processing into a concentrated form for food use, the cheese factory has had to install or utilize sanitary storage tanks and cooling equipment to prevent the outgrowth of food infectious or poisoning organisms prior to shipment. Washing and cleaning of such equipment would constitute another waste generating step not shown in the flow diagrams.

The cottage cheese operation differs from the other cheese processing in that it produces a whey of low pH (4.5 to 4.7) and utilizes washwater (equal to twice the whey volume) to remove residual whey, lactose and lactic acid from the curd. Cottage cheese curd is more fragile than rennet curd which is used for the other three types of cheese.

Thus, the whey and washwater may contain appreciable fine curd particles ("fines"). The amount of "fines" in the washwater is increased if mechanical washing (such as the Curd-O-Matic) processes are used. Complete draining of the curd, a prewash of 5% of the whey volume with draining, and a combination of this material with the whey to be sold for feed or food use or use of whey to cool the curd can decrease the BOD of the washwater by 20 to 40%.

In cheddar cheese, the whey draining during cheddaring (matting of the curd) and pressing should be collected and combined with the whey. In Swiss cheese manufacture, appreciable whey is lost in the transfer of the curd (150 to 300 lbs.) from the vat to the draining table. In provolone and mozzarella manufacture, the milling, mixing and molding of the curd produces a high fat, low pH (5.1 to 5.3) washwater from which 3 to 5% fat can be recovered by centrifugation.

The manufacture of cheese is one of the biggest aspects of dairy operations in the United States. Approximately one and one-quarter billion pounds of cheese are produced annually in the United States, and in the process approximately one and one-quarter billion pounds of whey solids are released. These whey solids are generated from the manufacture of both natural and cottage cheeses. A typical flow diagram for the manufacture of natural cheese by the enzyme coagulation method is shown in Figure 26.

FIGURE 26:   FLOW DIAGRAM OF CHEESE, LACTOSE, WHEY
PROCESSING OPERATIONS

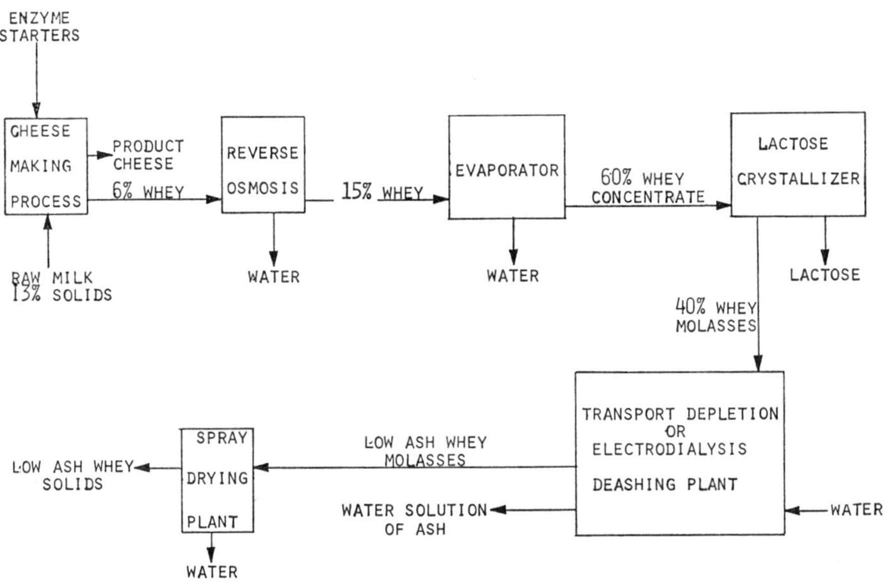

Source: OSW Publication No. 581

Cottage cheese is produced in a slightly different manner through the acid coagulation of milk. Whey production is divided so that approximately fourteen billion pounds of fluid are produced annually from natural cheese, while approximately half that quantity, or seven billion pounds per year, are produced from cottage cheese operations. Of the cheese marketed in the United States, approximately one-third is of the cottage cheese type, while the remaining two-thirds are of the natural or processed types.

The solids present in whey are a mixture of milk protein or lactalbumen, milk sugar or lactose, and a small amount of ash or dissolved salts. As whey is produced in the cheese making process, it is collected at a concentration of about 6% by weight. In other words, every 100 gallons of fluid whey contains approximately 50 pounds of whey solids. This quantity of whey solids is approximately equal to, or slightly less than, the weight of cheese which was extracted from the whole milk depending upon the cheese making process.

From the flow diagram of cheese making, the various steps of whey process-ing can be seen. Whole milk is either enzyme or acid coagulated to form the product cheese. The cheese curd is skimmed from the remaining fluid portion of the milk which is referred to as whey. The whey extracted from the cheese making vat is normally about 6% weight concentration. The whey solids themselves are composed of approximately 70% lactose, 20% lactalbumens, and 10% ash. Varying amounts of acids, ash, and other constituents may be present depending upon the specific mode of cheese making operation.

The first step in whey processing can be the preconcentration of whey from 6% solids to approximately 15% solids through the use of reverse osmosis. This step has been demonstrated to be feasible and is distinct in lower oper-ating costs over the conventional types of evaporating equipment. At about 15% concentration the osmotic pressure of whey solution becomes too high for continued reverse osmosis processing. At this point conven-tional milk or dairy evaporating equipment is used to concentrate the whey to approximately 60% solids.

Dairy evaporators are of a special sanitary, low temperature construction that is not found in the desalination industry. On this basis no potential is assumed for desalination type evaporators in this industry. At the 60% solids concentration level, lactose can be crystallized from the whey through cooling of a warm solution. About one-half of the lactose present is nor-mally removed in this manner. The remaining whey molasses at approxi-mately 50% concentration can be sent into a transport depletion or electro-dialysis deashing plant. In this equipment anywhere from 50 to 95% of the dissolved ionized salts can be removed resulting in a substantial upgrading of whey value.

The final step in whey molasses processing is typically the spray drying of the fluid to a stable dry powder. In this form the product has a long shelf life and can be easily transported and marketed. In the fluid form the whey molasses must be used within approximately twenty-four hours or bac-terial action may result in spoilage.

Data with respect to the exact location and quantity of production of cheese manufacturing plants throughout the United States is not available, or at least extremely difficult to compile. Data taken from Census Bureau and Department of Agriculture sources indicate that about 80% of the whey production is in the North Central United States.

The Census Bureau statistics for cheese production shown in Table 37 (production of whey, by weight, approximately equals production of cheese) indicates that on an average basis a typical cheese plant within the United States produces about one million pounds per year. This figure, it must of course be realized, reflects both the small private cheese producer and the ultralarge mechanized operations.

TABLE 37:   TOTAL CHEESE PRODUCTION

STATE	STATE TOTALS 1000 LBS. 1968	RANKING 1968	REGION TOTAL 1000 LBS. 1968
MID ATLANTIC	42,070		53,130
New York	42,070	6	
EAST NORTH CENTRAL	645,080		645,080
Ohio	12,690	18	
Indiana	18,580	13	
Illinois	13,640	17	
Michigan	22,410	11	
Wisconsin	577,760	1	
WEST NORTH CENTRAL	353,555		353,555
Minnesota	89,420	3	
Iowa	78,570	4	
Missouri	95,260	2	
North Dakota	24,735	10	
South Dakota	30,650	9	
Nebraska	15,360	15	
Kansas	19,560	12	
SOUTH	139,370		139,525
Arkansas	12,410	19	
Oklahoma	5,865	22	
Texas	9,230	21	
Kentucky	57,170	5	
Tennessee	36,390	7	
Alabama	4,540	24	
Mississippi	13,765	16	
MOUNTAIN	45,510		46,895
Montana	2,990	25	
Idaho	33,210	8	
Utah	9,310	20	
PACIFIC	25,705		25,707
Washington	5,675	23	
Oregon	17,760	14	
California	2,270	26	
TOTAL USA	1,251,290		1,263,892

Source: OSW Publication No. 581

Figure 27 shows a block flow diagram for cottage cheese manufacture.

FIGURE 27: FLOW DIAGRAM FOR COTTAGE CHEESE

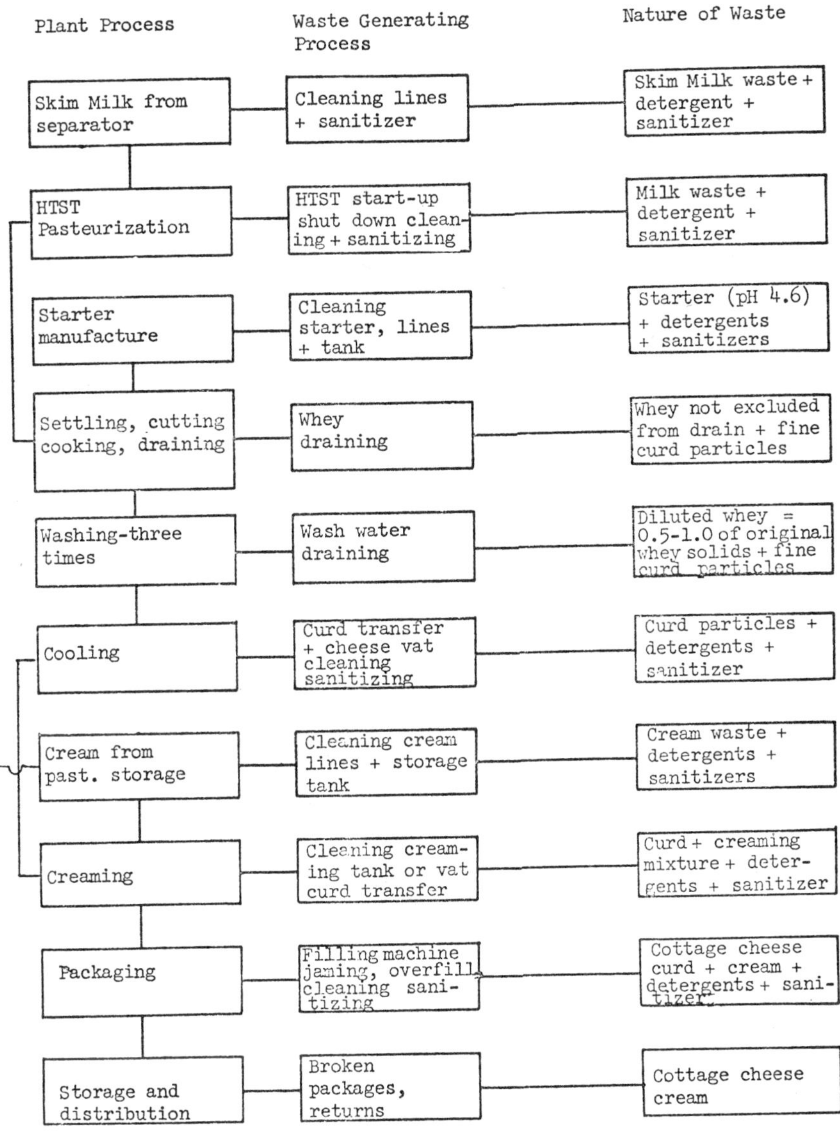

Source: EPA Report 12060 EGU, March 1971

Table 38 gives data in tabular form on the types of wastes from cottage cheese processing.

Figure 28 presents waste coefficients for cottage cheese manufacture.

TABLE 38:   TYPE OF WASTES FROM COTTAGE CHEESE PROCESSING

Plant Operation	Waste Generation From	Type of Waste
Pasteurized SM	See fluid milk products to transfer and storage	MS, DET, SAN
Starter manufacture	Cleaning lines and tank	Starter (pH 4.6) DET, SAN
Setting, cutting		
Cooking, draining	Whey draining	Whey (pH 4.6) Fines
Washing – 3 times	Wash water draining	Diluted whey, fines
Cooling	Curd transfer and vat cleaning and sanitizing	Curd particles and DET, SAN
Pasteurized C	Cleaning lines and tank	C, DET, SAN
Creaming	Clean tank or vat curd transfer	CURD, SAN, DET, Creaming mixture
Packaging	Filling Machine Jamming + Filling Cleaning	Creamed Cottage Cheese, DET, SAN
Storage and distribution	Broken packages returns	Creamed Cottage cheese

Source: Report PB 220,704

Figure 29 is a flow diagram for cheddar cheese manufacture and Figure 30 shows the waste coefficient for the various steps in cheddar cheese manufacture.

FIGURE 28:   WASTE COEFFICIENTS FOR COTTAGE CHEESE

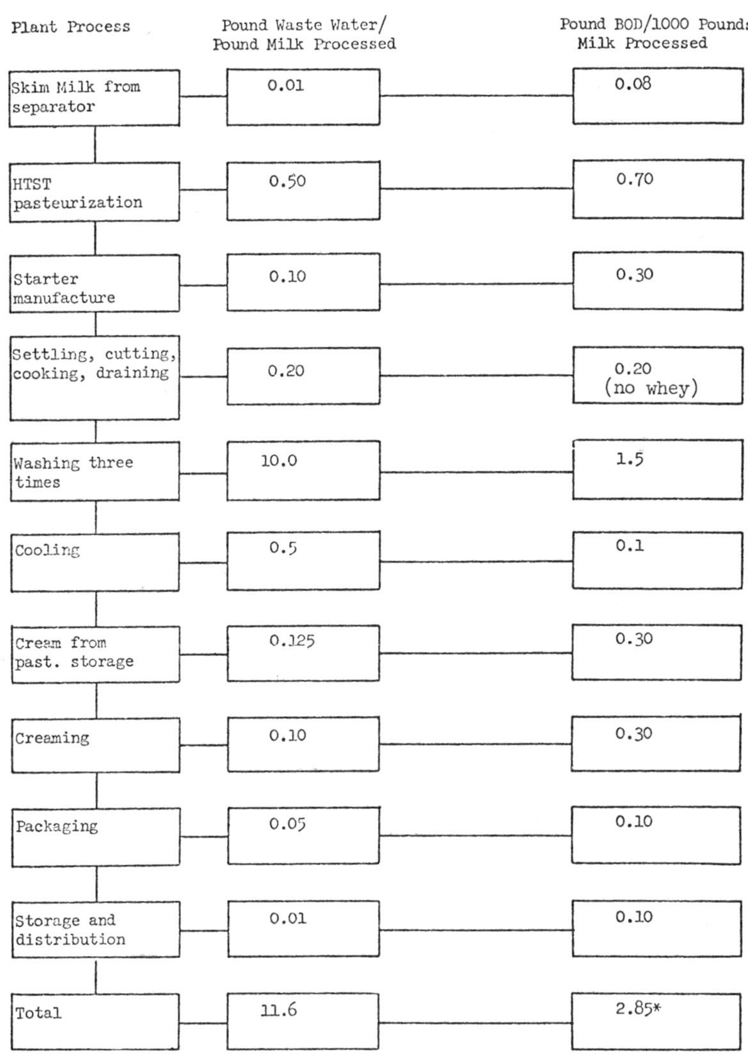

Source: EPA Report 12060 EGU, March 1971

FIGURE 29:   FLOW DIAGRAM FOR CHEDDAR CHEESE MANUFACTURE

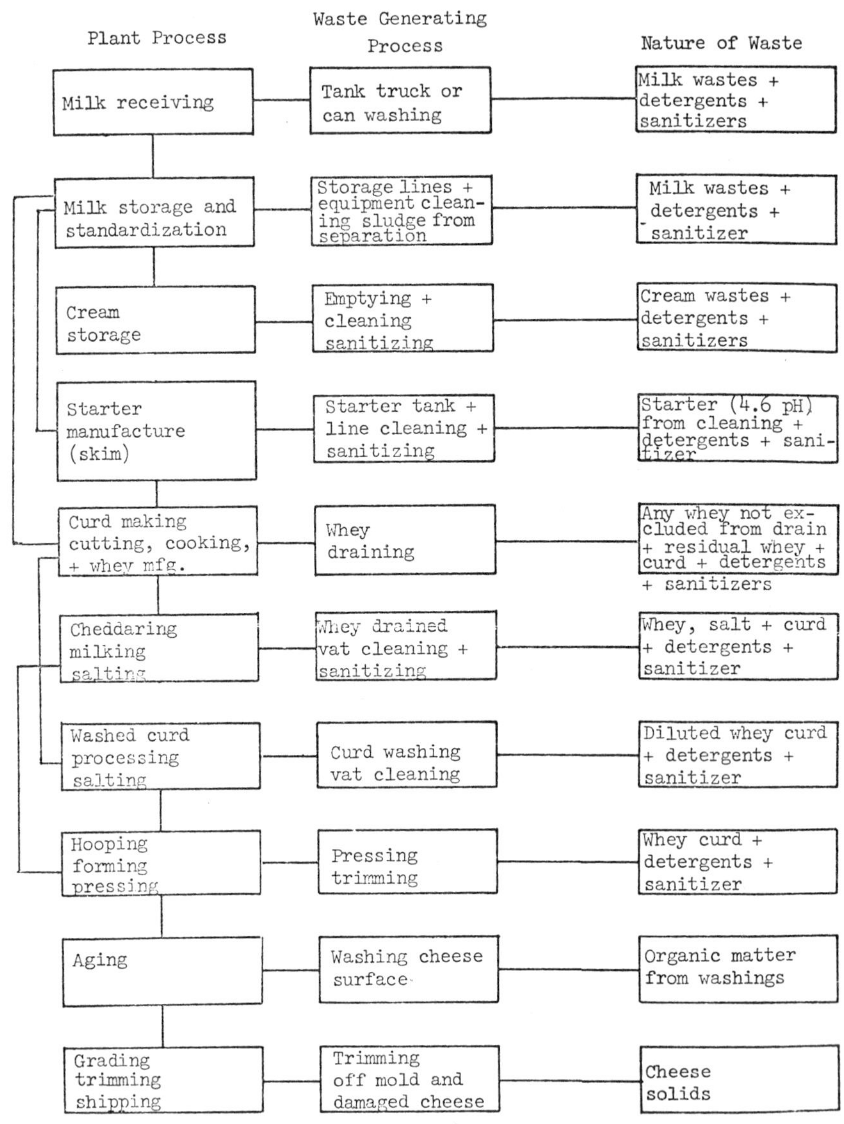

Source: EPA Report 12060 EGU, March 1971

FIGURE 30: WASTE COEFFICIENTS FOR CHEDDAR CHEESE MANU-FACTURE

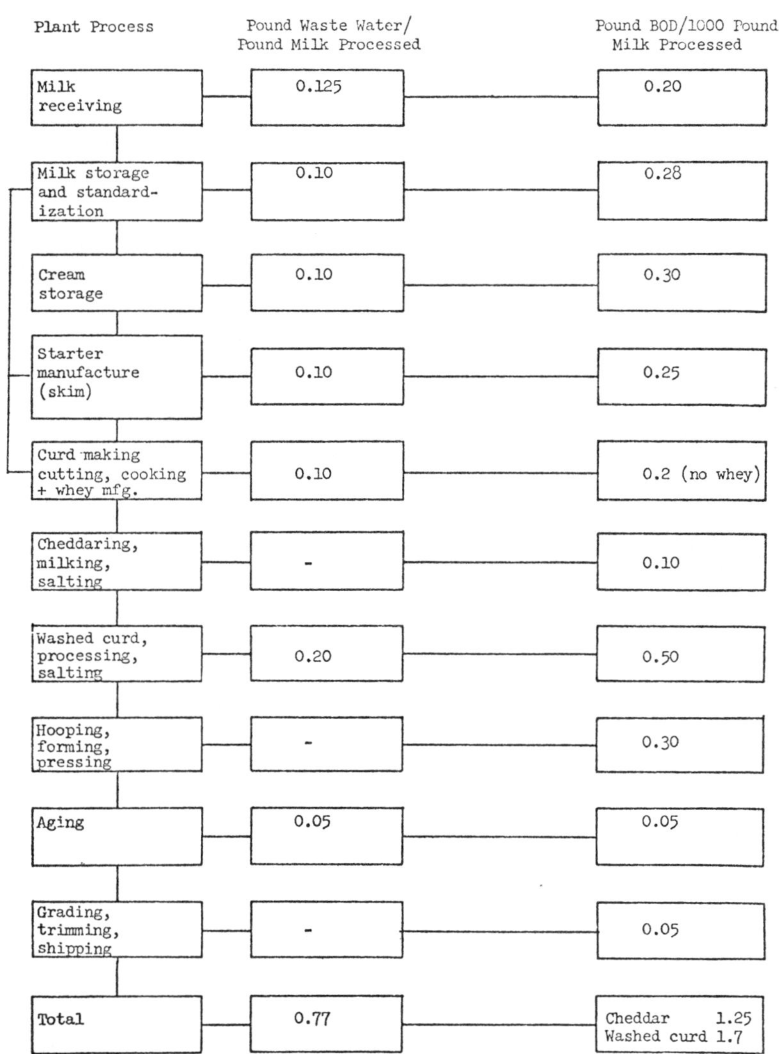

Source: EPA Report 12060 EGU, March 1971

Figure 31 is a block flow diagram for Swiss cheese manufacture.

FIGURE 31:   FLOW DIAGRAM FOR SWISS CHEESE MANUFACTURE

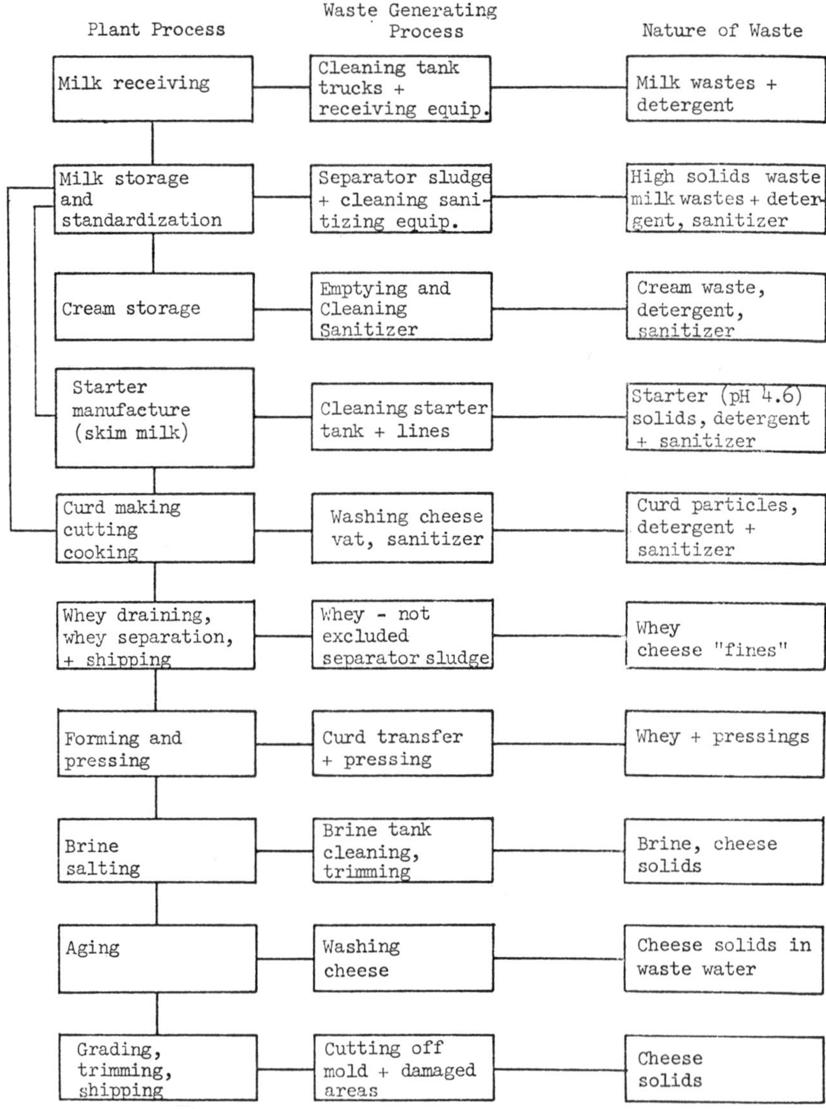

Source: EPA Report 12060 EGU, March 1971

Figure 32 is a block flow diagram for the manufacture of cheese pasta filata and Figure 33 shows the corresponding waste coefficients for the various process steps.

FIGURE 32:   FLOW DIAGRAM FOR PASTA FILATA CHEESE MANU-
FACTURE

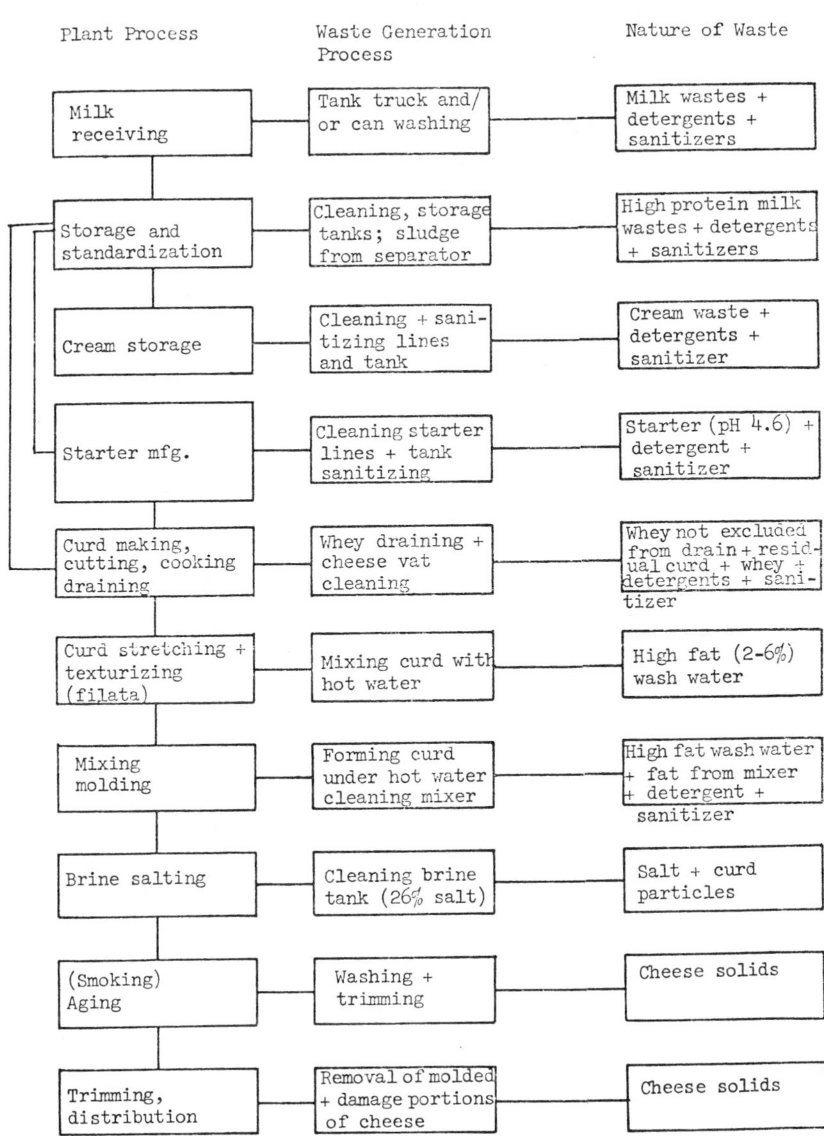

Source: EPA Report 12060 EGU, March 1971

FIGURE 33: WASTE COEFFICIENTS FOR PASTA FILATA CHEESE MANU-
FACTURE

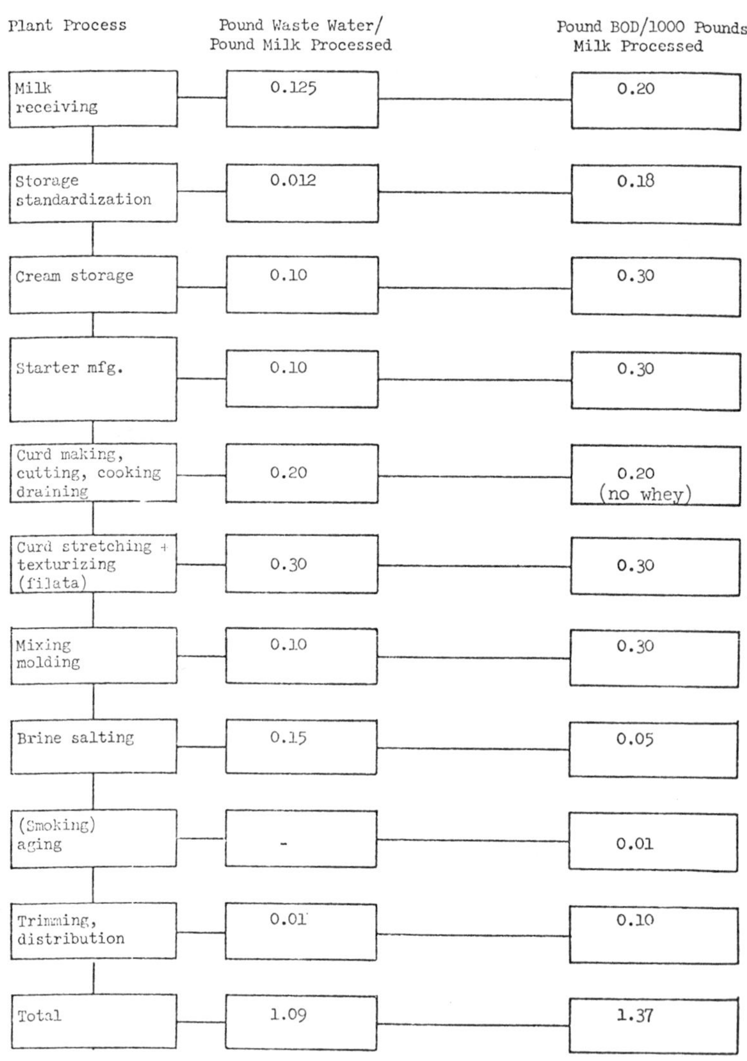

Source: EPA Report 12060 EGU, March 1971

Because of its significance, the cottage cheese operation merits special attention. Cottage cheese whey BOD_5 values in industrial practice were found to range from 30,000 ppm for filtered whey to 65,000 ppm for whey with a high level of fine particles present. Thus, in addition to whey and washwater, the amount of curd shattering can be a significant source of both BOD and suspended solids in cottage cheese operations. The BOD_5 for whey, washwater and fines in one commercial dairy food plant is presented in Table 39. As indicated, the total BOD value of the fines was lbs./1,000 lbs. of milk processed into cottage cheese. The first washwater contained about 60% of the total BOD in the washwaters.

Effort is needed to minimize the BOD in washwaters, since the solids level (less than 2.0%) is too low for practical recovery. This may be accomplished by: (a) complete draining of the curd, followed by addition of 10% volume of washwater that is drained and combined with the whey, before the first complete washing, (b) utilization of whey as a means of cooling the curd, in which whey is removed, cooled and pumped back on the curd, thus reducing the volume of washwater required, and (c) countercurrent washing through a cooling tower. The possibility exists also for developing new methods of cottage cheese manufacture from concentrated milk which would reduce the volume of whey and washwater, but this is in the future from a practical viewpoint.

TABLE 39: BOD COEFFICIENTS FOR COTTAGE CHEESE OPERATIONS*

Operation	ppm BOD		#BOD/1000 lb. milk proc.	
	A	B	A	B
Whey C. C.	3,700	3,500	26.6	25.2
First wash	17,000	11,700	5.8	6.3
Second wash	1,700	2,200	1.3	1.2
Third wash	2,000	550	0.7	0.30
Fourth wash	2,200	--	0.75	--
Transfering & Filling+	--	185,000	--	0.027
Fines	--	--	--	2.2

*Operation A used 3 washings with each wash being equal to 75% of whey volume. Operation B used 5 washes, with the 1st, 3rd, and 4th washes being equal to 50% of the whey volume and the 2nd wash being equal to the whey volume.

+Values based on curd spillage. Usual amount of curd loss would range from 5 to 10 lbs./100 lbs. of cottage cheese.

Source: Second National Symposium on Food Processing Wastes (1971)

Data in Table 40 show the BOD coefficients for two cottage cheese operations based on information concerning the BOD values of various cheese washwaters. The concentration, percentage volume of original whey, and BOD of whey washwater and fines available from one industrial plant is presented in Table 41.

These data represent values from four vats. The values are in relatively good agreement with the values in the preceding table and further illustrate the significance of solid casein particles on BOD and the necessity that these materials be excluded from the operations. Careful draining and a partial prerinse can reduce the BOD to approximately 50% of the values shown. The pounds of BOD in the washwater are about 33% of the pounds of BOD in the whey.

TABLE 40:   BOD COEFFICIENTS FOR COTTAGE CHEESE OPERATIONS*

Operation	ppm BOD		Lb BOD/1000 Lb		Lb BOD/1000 Lb Milk Processed	
	A	B	A	B	A	B
Whey, Cottage Cheese	3,700	35,000	37	35.0	26.6	25.2
First wash	17,000	11,700	17	11.7	5.8	6.3
Second wash	1,700	2,200	1.7	2.2	1.3	1.2
Third wash	2,000	550	0.4	0.55	0.7	0.30
Fourth wash	2,200	--	0.44	--	0.75	--
Transferring and Filling**	--	185,000	--	185.0	--	0.027

*Operation A used three washings, each wash being equal to 75% of whey volume. Operation B used four washes, with the first, third and fourth washes being equal to 50% of the whey volume and the second wash being equal to the whey volume.

**Values based on curd spillage. Usual amount of curd loss would range from 5-10 pounds/100 pounds of cottage cheese.

Source: EPA Report 12060 EGU, March 1971

Over the years, industry has utilized whey as a means of cooling curd instead of washwater, thus minimizing the amount of whey required. Although this was done originally to avoid contamination of the cheese with spoilage microorganisms, the same can be accomplished today to reduce both the volume and BOD strength of washwater.

The practice was to pump the whey into the vat, put it through a heat exchanger in two stages to bring the temperature of the curd down to about 50°F. The curd was then thoroughly drained and washed with a volume equal to one-half of that of whey as a final wash to reduce acid flavor in the cheese. Utilization of such a procedure could reduce the BOD level in washwater diluted solids down to about 50% of the current levels.

The care taken in manufacture to avoid curd breakage is significant in respect to control of the BOD in the wastewater. Consideration needs to be given to the possibility of developing cheese making processes which would basically eliminate whey or minimize it by using concentrated milk as a starting material or the development of an entirely new process based on dry milk solids.

TABLE 41:   BOD COEFFICIENTS FOR WHEY AND WASHWATER FROM COTTAGE CHEESE MANUFACTURE

Operation	Volume as Percent of Whey	Percent Solids in Whey		Percent Fines		BOD of Whey/ 1000 pound Wash Water Ave.	BOD of Whey ppm, Ave.	BOD of Fines, Ave.
		Range	Ave.	Range	Ave.			
Whey	100%	6.47-6.77	6.61	0.13-0.20	0.16	44.8	44,800	1.6
1st wash	75	1.23-4.22	3.24	0.04-0.21	0.11	21.3	21,300	1.1
2nd wash	75	0.48-1.18	1.21	0.04-0.37	0.13	2.46	4600	1.3
3rd wash	75	0.16-0.33	0.26	0.0 -0.2	0.01	3.2	3200	0.01

Source: EPA Report 12060 EGU, March 1971

ICE CREAM OPERATIONS

The unit operations, sources of wastes and waste coefficients are indicated in the figures which follow for ice cream, frozen dessert operations and subprocesses. Alternative pathways are illustrated for novelty preparations. Because of high viscosity and high BOD values, incomplete draining of ice cream and ice cream mix lines and equipment can be a major source of BOD and suspended solids (SS) (nuts, fruits, etc.) in dairy food plant wastewater.

Figure 34 is the block flow diagram for ice cream manufacture.

FIGURE 34:   FLOW DIAGRAM FOR ICE CREAM

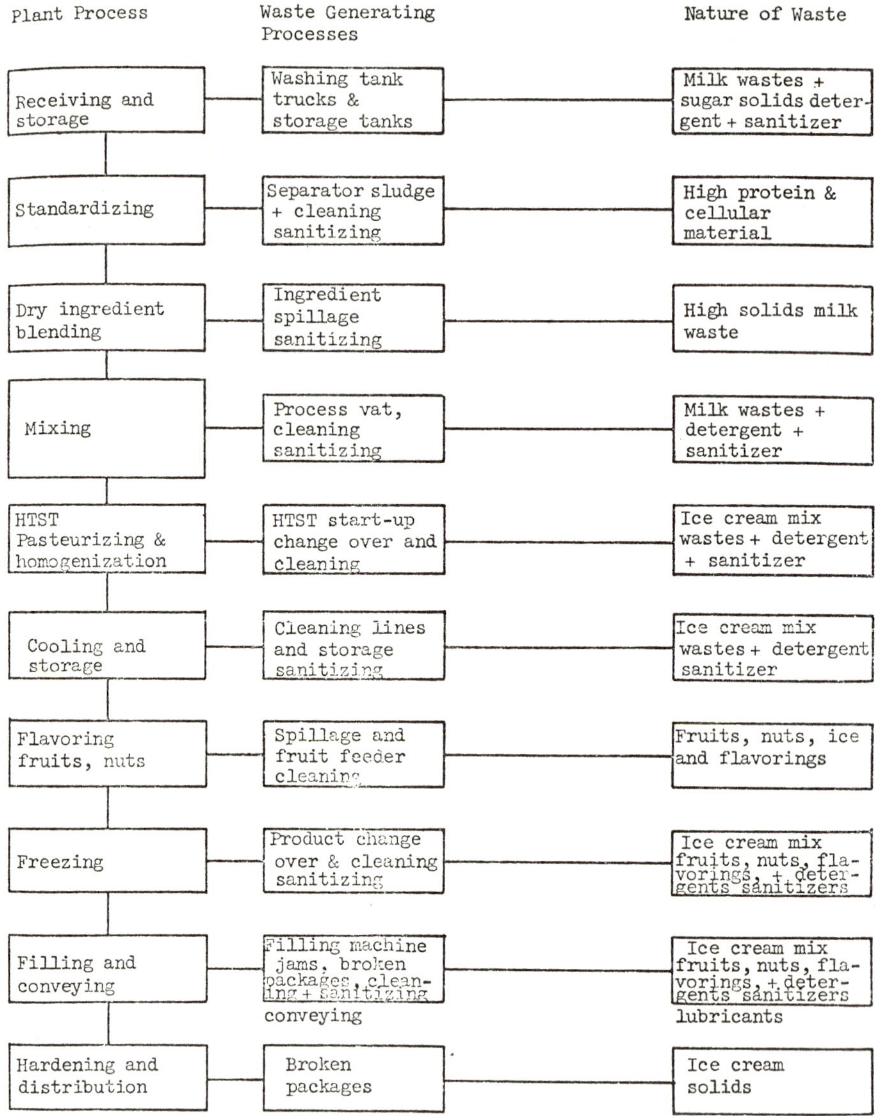

Source: EPA Report 12060 EGU, March 1971

In processing of frozen products, the waste included cleaning compounds
and sanitizers in addition to ingredients. It is especially important to
minimize product losses in the frozen products area because of the relatively high solids content of these products. The concentration of ice
cream and ice milk mix make for high costs and large waste loads if product escapes.

Considerable product is lost in frequent changeovers and in cleanup due
to the viscous nature of the products. Both low temperature and high
solids content contribute to the viscosity. Product in vats, lines, freezers
and other processing and packaging equipment does not drain as fully as
less viscous products. Mix left in the lines usually ends up as waste.
Table 42 shows some of the types of wastes resulting from ice cream processing.

TABLE 42:   TYPES OF WASTES FROM ICE CREAM PROCESSING

Plant Operation	Waste Generation From	Type of Waste
Receiving and storage	Washing storage tanks	MS, DET, SAN, Sugars
Dry ingredient blending	Ingredient spillage sanitizing	MS (high solids)
Mixing	Process vat, cleaning sanitizing	MS, DET, SAN
HTST	HTST start-up, change over and cleaning	SAN, DET ICM wastes
Cooling and storage	Cleaning lines and storage tanks, sanitizing	ICM wastes, DET, SAN
Flavoring fruits nuts	Spillage and fruit feeder cleaning and cleaning flavoring vat	ICM, Fruits, nuts, flavorings, DET, SAN
Freezing	Product change over and cleaning, sanitizing, freeze up	ICM, Fruits, nuts, flavorings, DET, SAN
Filling and conveying	Filling machine jams, broken packages, cleaning sanitizing, conveying, Product change over	ICM fruits, nuts, flavorings DET, SAN, LUB
Hardening and distribution	Broken packages	Ice Cream Solids

Source: Report PB 220,704

Figure 35 shows the waste coefficients for the various steps in ice cream manufacture.

FIGURE 35:   WASTE COEFFICIENTS FOR ICE CREAM

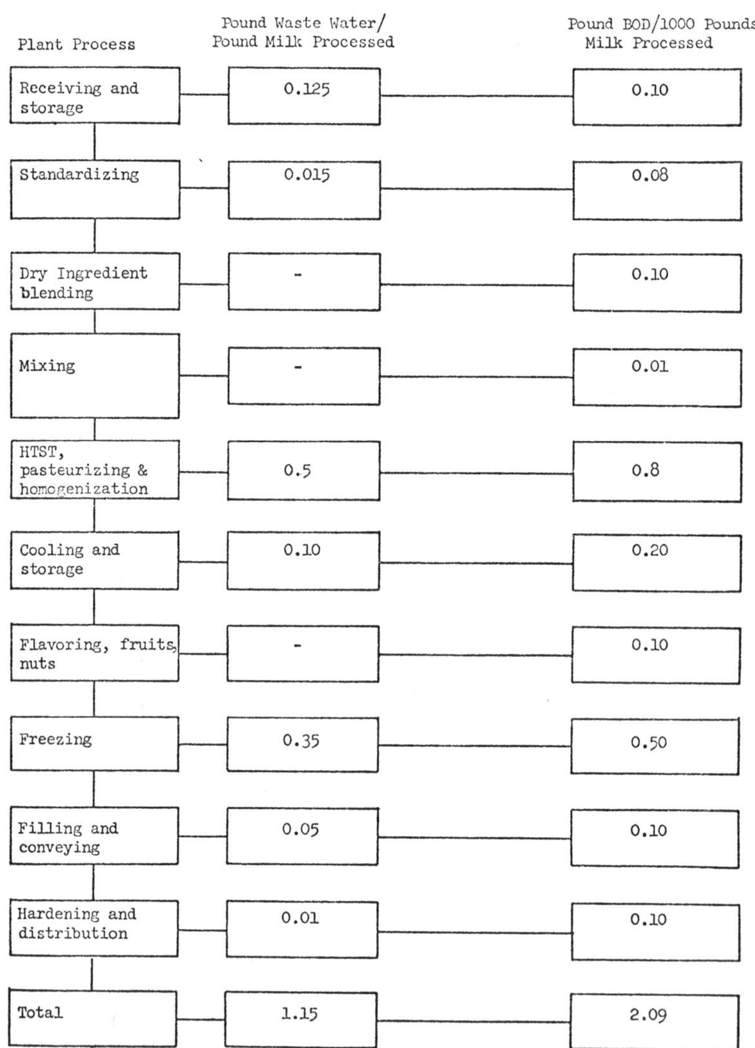

Source: EPA Report 12060 EGU, March 1971

Figure 36 is a block flow diagram for ice cream stick novelties manufacture. Table 43 itemizes the types of wastes produced by ice cream stick novelty processing and Figure 37 shows the waste coefficients for the various steps in the process.

FIGURE 36: FLOW DIAGRAM FOR ICE CREAM SUBPROCESS — STICK NOVELTIES

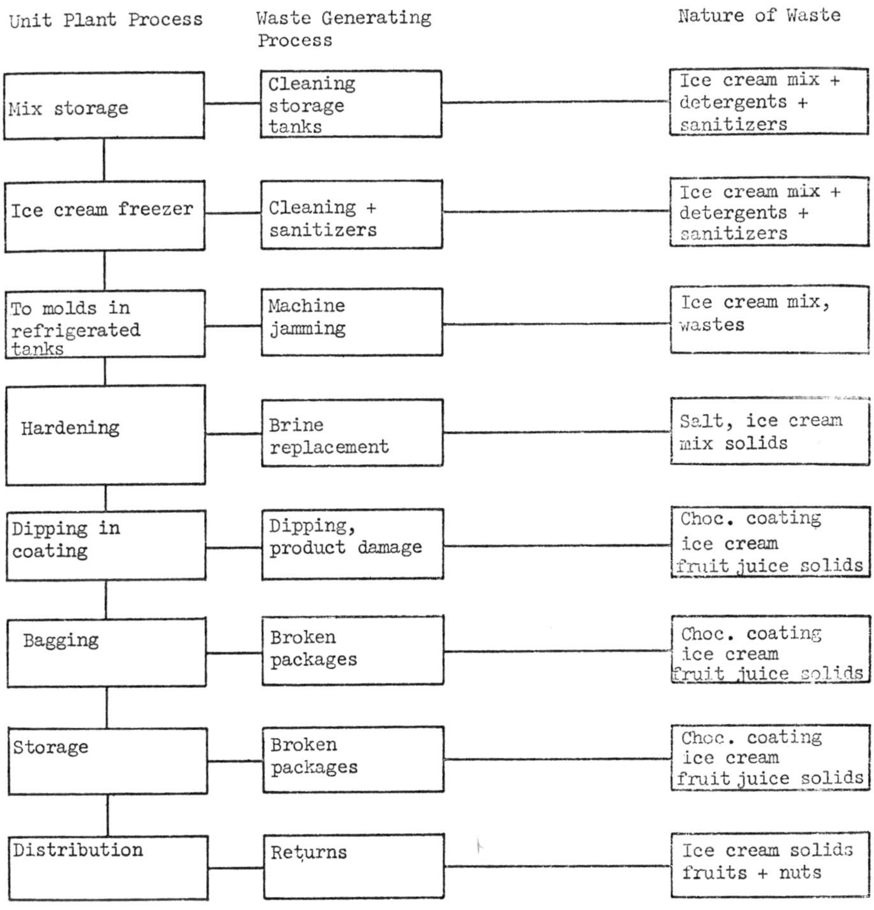

Source: EPA Report 12060 EGU, March 1971

FIGURE 37:   WASTE COEFFICIENTS FOR ICE CREAM SUBPROCESS —
STICK NOVELTIES

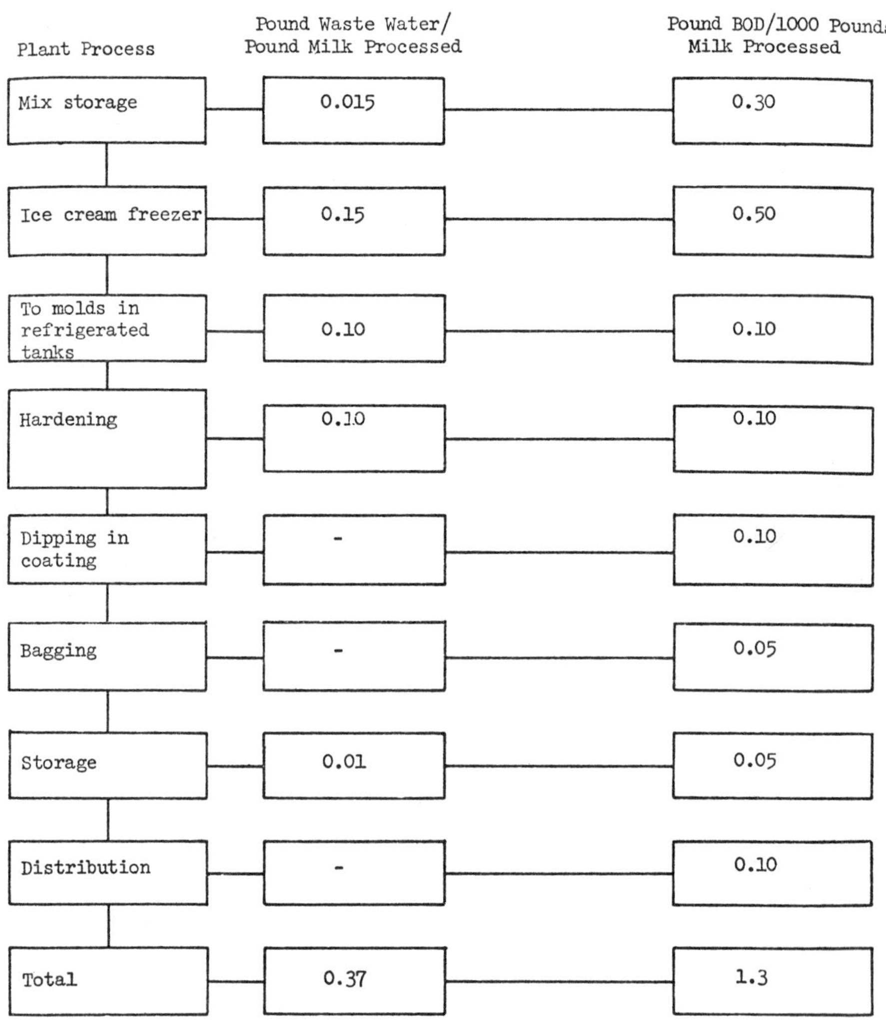

Source: EPA Report 12060 EGU, March 1971

Figure 38 is a block flow diagram for ice cream stickless novelties manufacture. Table 44 itemizes some of the wastes produced by stickless novelty
manufacture and Figure 39 shows the waste coefficients for the various
steps in stickless novelty manufacture.

FIGURE 38: FLOW DIAGRAM FOR ICE CREAM SUBPROCESS — STICKLESS NOVELTIES

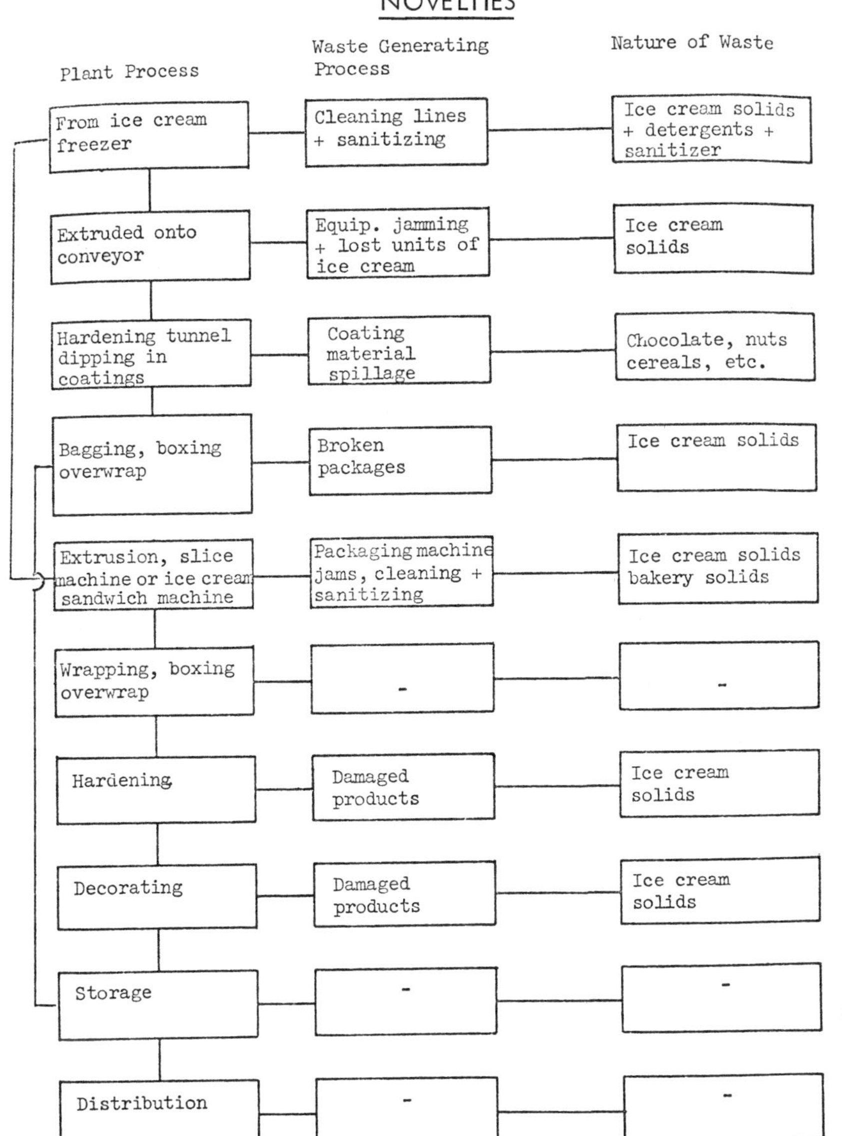

Source: EPA Report 12060 EGU, March 1971

TABLE 43:   TYPE OF WASTES FROM STICK NOVELTY PROCESSING

Plant Operation	Waste Generation From	Type of Waste
Mix storage	Cleaning storage tanks	DET, ICM, SAN
Ice cream freezer	Cleaning and sanitizers	ICM, DET, SAN
Mold filling	Machine jamming, overfilling	ICM
Hardening	Brine replacement	Salt, ICM
Dipping in coating	Dipping, product damage	Choc. coating ICM, fruit juice solids, nuts
Bagging	Broken packages	Choc. coating ICM, fruit juice solids
Storage	Broken packages	Choc. coating, ICM, fruit juice solids
Distribution	Returns	Ice cream solids fruits and nuts

TABLE 44:   TYPE OF WASTES FROM ICE CREAM NOVELTY PROCESSING

Plant Operation	Waste Generation From	Type of Waste
From ice cream freezer	Cleaning lines, sanitizing	ICM, DET, SAN
Extruded onto conveyor	Equip. jamming, lost units of ice cream	ICM
Hardening tunnel dipping in coatings	Coating material spillage	Chocolate, nuts, cereals, etc.
Bagging, boxing overwrap	Broken packages	ICM
Storage, distribution		
From ice cream freezer	Cleaning lines, sanitizing	ICM, DET, SAN
Extrusion, slice machine or ice cream sandwich machine	Packaging machine jams, cleaning, sanitizing	ICM Bakery solids
Wrapping, boxing overwrap		
Hardening	Damaged products	ICM
Decorating	Damaged products	ICM
Storage, distribution		

Source: Report PB 220,704

FIGURE 39:   WASTE COEFFICIENTS FOR ICE CREAM SUBPROCESS — STICKLESS NOVELTIES

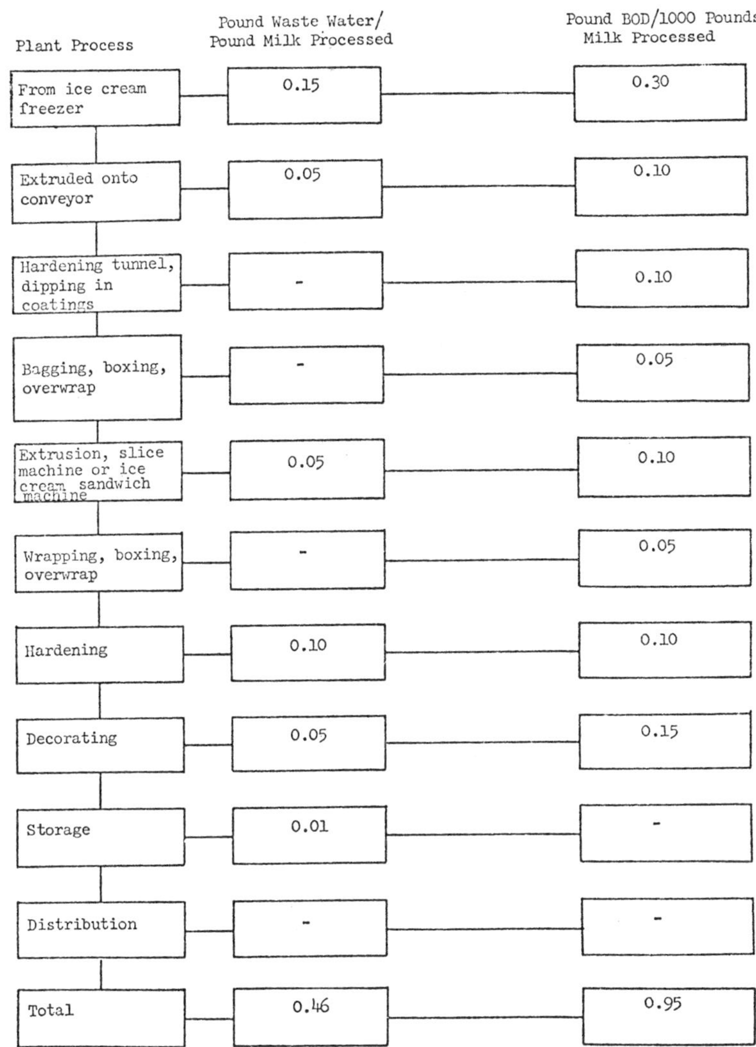

Source: EPA Report 12060 EGU, March 1971

CONDENSED AND EVAPORATED MILK PRODUCTION

Figure 40 is a block flow diagram of the condensed milk production process showing points of waste generation. Figure 41 shows the waste coefficients for the condensed milk process.

FIGURE 40:   FLOW DIAGRAM FOR CONDENSED MILK PROCESS

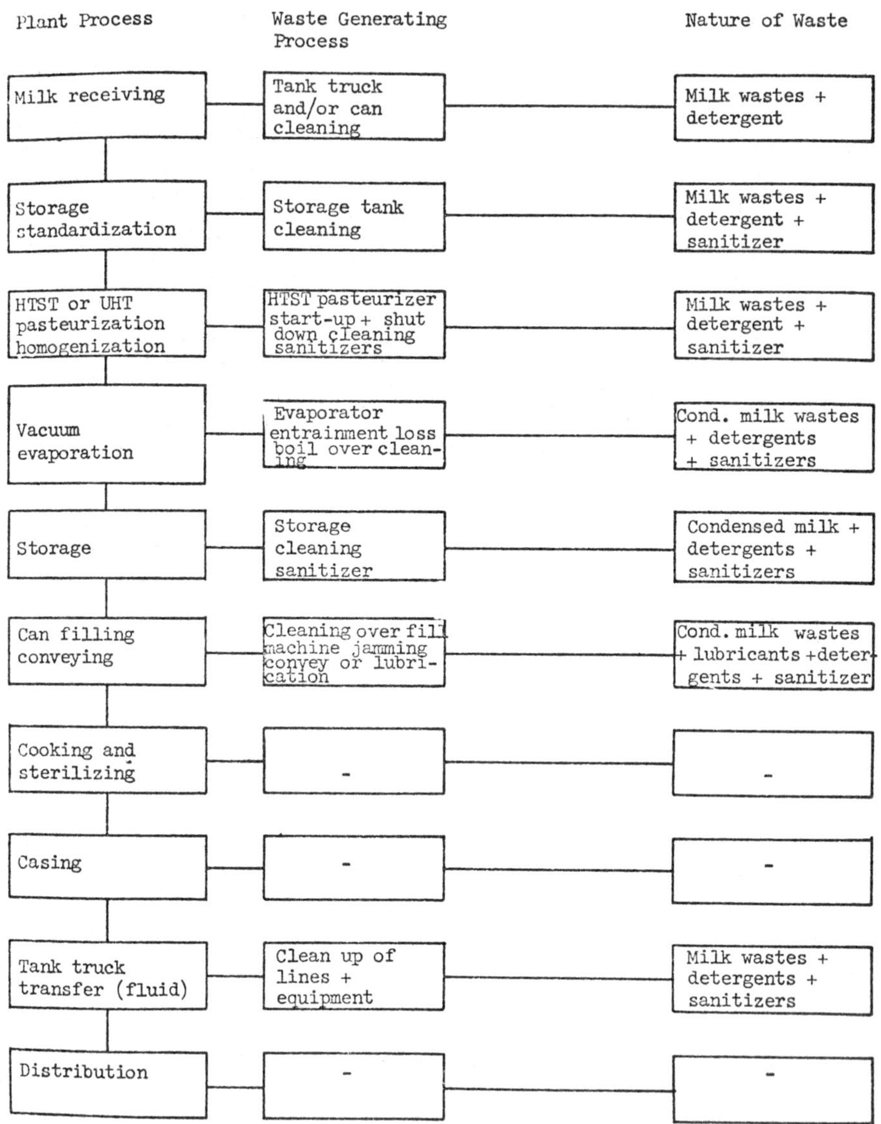

Source: EPA Report 12060 EGU, March 1971

FIGURE 41:   WASTE COEFFICIENTS FOR CONDENSED MILK PROCESS

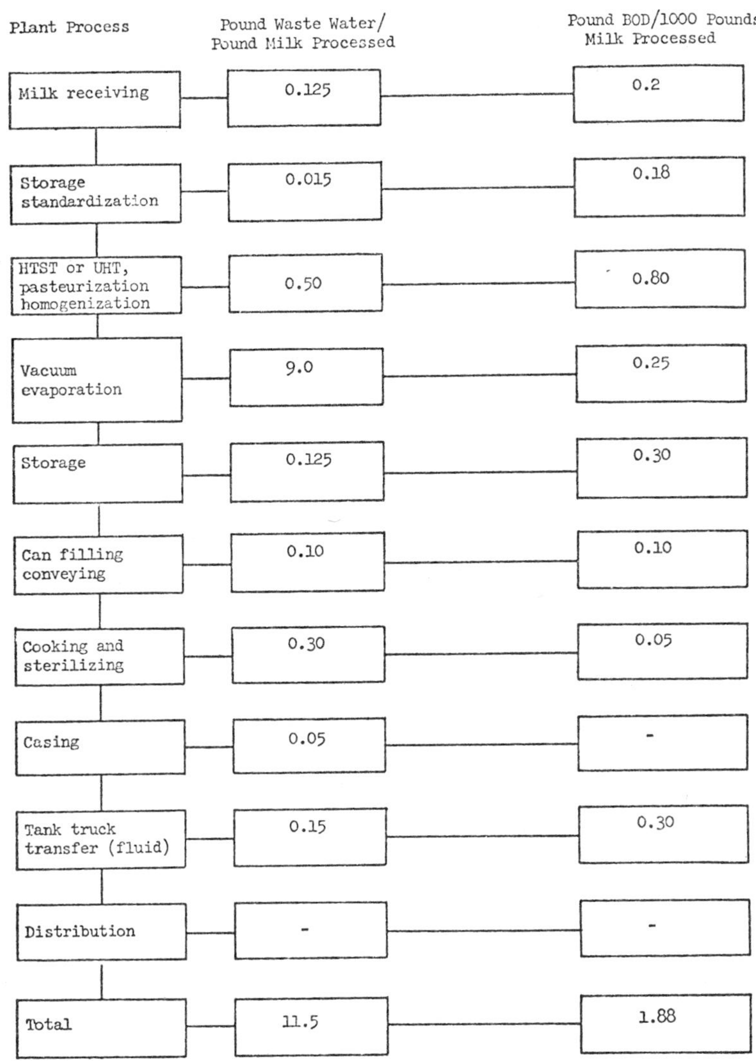

Source: EPA Report 12060 EGU, March 1971

Figure 42 is a block flow diagram of the spray drying process for evaporated milk production featuring waste origins and Figure 43 shows the waste co-efficients for the spray drying process.

FIGURE 42:   FLOW DIAGRAM FOR SPRAY DRYING PROCESS

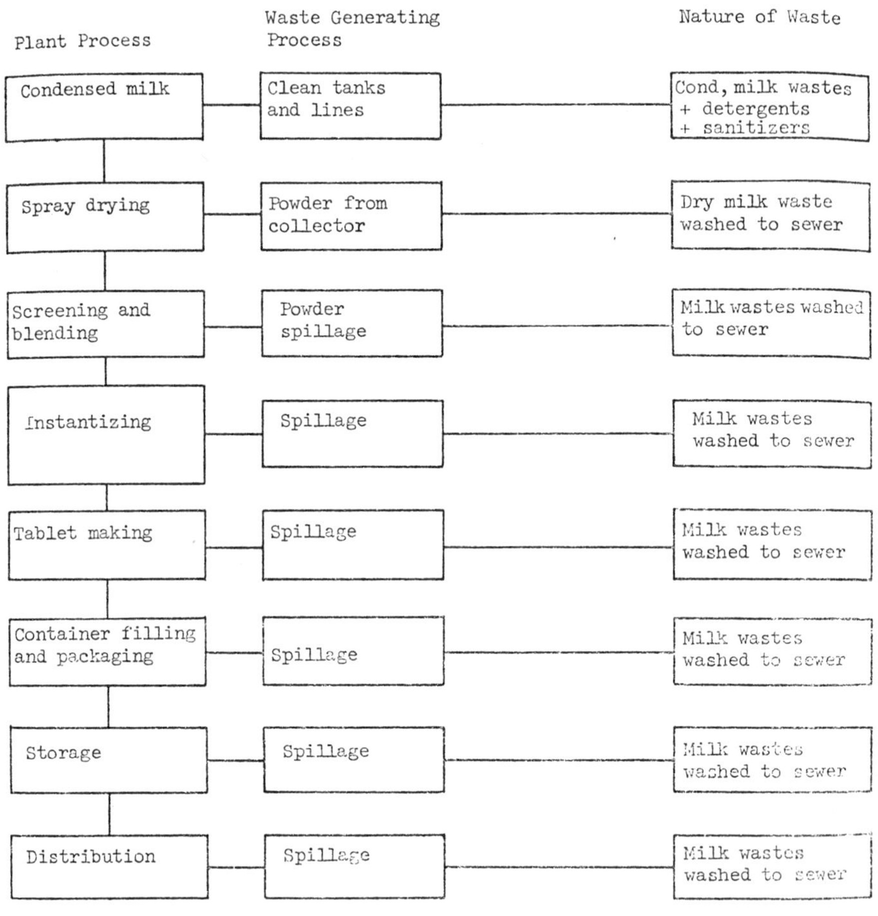

Source: EPA Report 12060 EGU, March 1971

FIGURE 43: WASTE COEFFICIENTS FOR SPRAY DRYING PROCESS

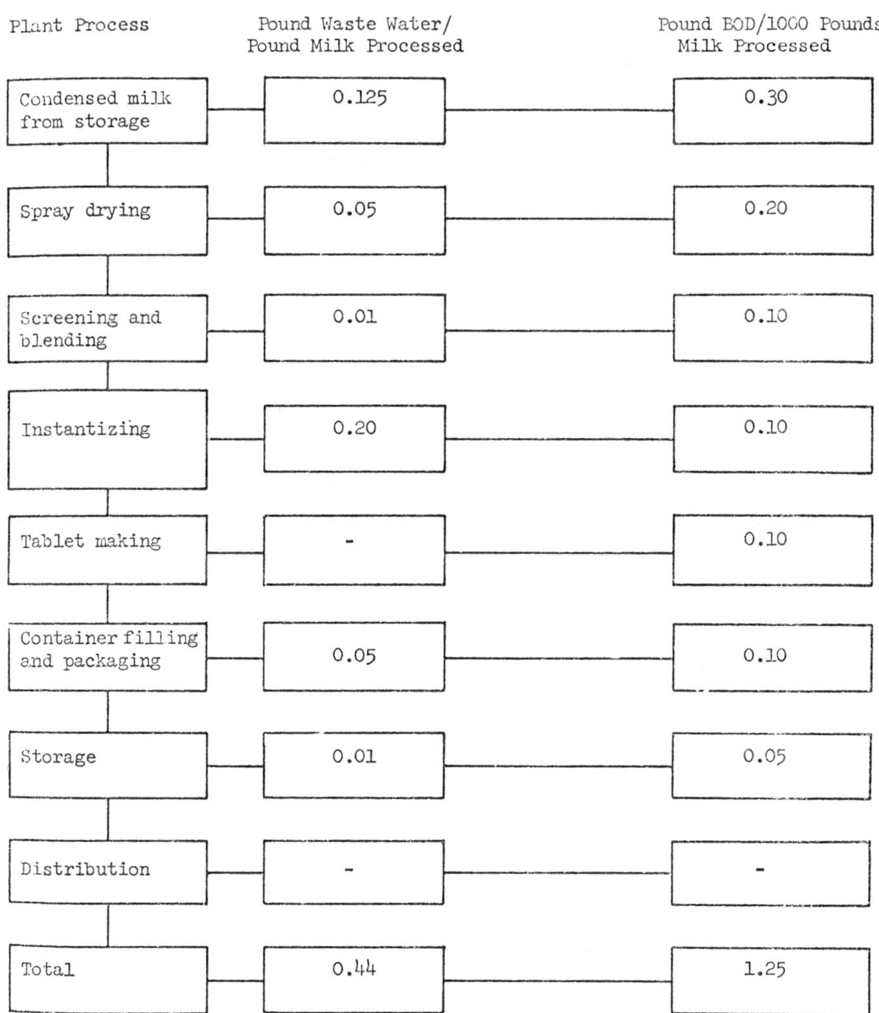

Source: EPA Report 12060 EGU, March 1971

IN-PLANT CONTROL TO REDUCE WASTES

A prevalent view within the dairy industry has been that the elimination of whey from dairy plant wastewaters would essentially solve the problem of fluid pollution for the dairy industry. Site visitations and information on waste coefficients obtained from industry would indicate that the elimination of whey does not necessarily solve the pollution problem.

In plants handling cottage cheese, where whey was excluded from the wastewater, BOD coefficients ranged from 0.7 to 9.6. The controlling factor in respect to in-plant control of dairy plant wastes appears to be management's attitude and the ability of middle management to properly supervise plant operations.

Control of waste coefficients in a dairy plant can take several pathways. The first includes those changes that can be put into operation without additional expense. The second group are the engineering changes that can affect material reduction in either wastewater volume or BOD.

In the course of the study on the State of the Art of Dairy Wastes and Dairy Waste Treatment (2), site visitations were made to 20 dairy plants for which waste coefficient data were available. It was possible to obtain an evaluation of management practices in some of the other plants for which waste coefficient data also were obtained. In evaluating management practices, the following criteria were utilized:

1. Housekeeping practices.

2. Water control practices; frequency with which hoses and other sources of water were left running when not in actual use.

3. Degree of supervision of operations contributing to either volume or BOD coefficients.

4. Extent of spillage, pipeline leaks, valve leaks and pump-seal leaks.

5. Extent of carton breakage and product damage in casing, stacking and cooler operations.

6. Practices utilized in handling whey.

7. Practices utilized in handling spilled curd particles during cottage cheese transfer and/or filling operations.

8. Utilization of practices to reduce the amount of wash water from cottage cheese or butter operations.

9. Extent to which the plant is utilizing procedures to segregate and recover milk solids in the form of rinses and/or products from pasteurization start-up and product change-over.

10. The procedures utilized in handling returned products.

11. Evaluation of the management attitude toward waste control.

Although the evaluations are subjective in nature and do not lend themselves to absolute quantitation, they do provide a valid impression of management overall quality of practices in respect to waste control.

The influence of management practices affecting waste coefficients, volume and BOD coefficients is summarized in Table 45 on the following page. Evaluation of the information provides a recognition that there is a direct relationship between the quality of management practices in respect to waste control and the waste coefficient. Under good control, both volume and BOD coefficients are low.

Under extremely good management, coefficients of 0.5 pound of wastewater per pound of milk processed and 0.5 pound of BOD per thousand pounds of milk processed were obtainable. A realistic average of 2.0 pounds of BOD per thousand pounds of milk processed and 1.5 pounds of wastewater per pound of milk processed appeared to be generally achievable under good management. Coefficients above 3.0 can be considered to be excessive and an indication of poor management. With whey and water milk excluded, the waste coefficients of different operations are quite similar.

TABLE 45: THE EFFECT OF MANAGEMENT PRACTICES ON WASTE COEFFICIENTS

Plant No.	Products Manufactured	Milk Processed #/day	#BOD/1,000# Milk Processed	#Wastewater/# Milk Processed	Level Management Practices	Explanation of Practices
1	Milk	400,000	0.3	0.4	Excellent	Rinses saved, hoses off – out of use, filler drip pans
42	Milk	150,000	7.8	5.2	Poor	No steps taken to reduce waste
43	Milk		0.2	0.1	Excellent	Rinses saved, returns excluded, filler drip pans, cooling power
6	Cottage cheese	600,000	2.0	0.8	Good	Whey excluded, fines screened out, wash water to drain.
36	Cottage cheese	300,000	1.3	4.7	Good	Whey excluded, spilled curd handled as solid waste
37	Cottage cheese	650,000	71	12.4	Poor	Whey included, poor housekeeping
9	Ice cream	17,000	32.2	5.3	Poor	Rinses to drain, leaks, drips, water running — not in use
26	Ice cream	34,000	2.1	0.8	Good	Freezer rinses segregated
48	Milk Cottage cheese	250,000	0.7	1.0	Good	Whey and wash water excluded, rinses segregated
8	Milk Cottage cheese	1,000,000	8.6	7.0	Poor	Whey excluded, many drips, leaks, returns included
10	Milk	900,000	3.3	1.1	Fair	Whey excluded, good water volume control
40	Milk Cottage cheese	1,000,000	2.1	1.2	Good	Whey excluded, rinses saved
52	Milk Cottage cheese	765,000	1.8	1.1	Good	Returns excluded, good water control
3	Milk Ice cream Cottage cheese	400,000	3.9	1.4	Fair	Whey and wash water excluded, rinses excluded
30	Milk Ice cream Cottage cheese	800,000	7.7	3.5	Poor to Fair	Whey excluded, sloppy operation — spillage, leaks, hoses running
33	Milk Ice cream Cottage cheese	600,000	12.9	3.3	Poor	Whey included; poor housekeeping — pumps, valves, lines leaking, hoses running
34	Milk Ice cream Cottage cheese	900,000	9.1	2.8	Poor	Whey excluded, many leaks, drips, etc.
44	Milk Butter	300,000	0.9	0.8	Good	Buttermilk excluded, few leaks, dry floor condition
50	Whey powder	500,000	0.2	5.9	Good to Fair	No entrainment losses, all powder handled as solid waste, no leaks or drips
56	Milk powder Butter	200,000	3.0	2.5	Fair	Continuous churn hoses running, numerous leaks and drips

Source: EPA Report 12060 EGU, March 1971

If the overall average coefficients of 2.4 for volume and 5.8 for BOD are assumed to be representative of the national dairy food industyr, the total wastewater from dairy food plant operations would be equivalent to 200 million pounds of wastewater with 475 million pounds of BOD_5 per year. Since observations have indicated that plants without prior knowledge of their wasteloads have coefficients in excess of three pounds of wastewater per pound of product processed and 3.0 pounds of BOD per thousand pounds of product processed, the actual amount of BOD from dairy food plants, considering a 50% by-product waste, an overall organic wasteload from dairy food of 4.0 billion pounds of BOD_5 can be estimated at this time.

REPORTED WASTES VS. WASTES CALCULATED FROM BOD5 VALUES

All dairy plants maintain records of incoming and outgoing milk and milk products and calculate a percentage of fat and nonfat solids loss on a monthly basis. The industry has frequently used these figures to predict the BOD values in the industry's wastewater. Where data was available from industry, reported losses were used to calculate probable BOD from milk solids sources and the reported BOD was used to back-calculate the percentage loss of whole milk from BOD going to sewer. In all cases the assumption was made that 10.0% of the BOD was derived from nonmilk solids origin, based on actual data from two plants. Thus 1.0 pound of BOD was converted to 9.0 pounds of 3.5% milk. Table 46 shows this data.

TABLE 46:   REPORTED AND CALCULATED MILK SOLIDS LOSSES

Plant No.	% Plant Loss Reported		Calculated BOD (lb) from Reported Loss	Reported Lb. BOD	% Loss of Whole Milk from Reported BOD
	% Nonfat Solids	% Fat			
1	0.35	0.38	188	139.20	0.278
2	0.68	1.28	1210	2401.20	2.030
3	0.14	0.83	450	1102.0	2.030
4	0.02	1.10	497	2818.80	6.264
6	--	1.0	805	1450.00	1.740
7	1.0	1.0	801	1763.20	2.204
9	0.16	0.14	5	747.04	9.960
10	0.5	0.5	543	3526.40	3.248
11	0.7	0.5	129	945.40	3.712
12	0.6	0.7	156	1067.20	4.756
18	0.87	0.60	156	963.96	3.596
25	0.09	0.09	4	83.52	1.856
26	1.4	1.5	33	634.52	28.884
28	--	0.35	169	1038.20	2.088
29	0.45	0.35	64.3	380.48	1.740
30	0.75	0.47	486	7115.44	6.844
34	1.0	1.4	1232	7900	8.9
36	1.1	1.1	33	55	1.8
37	0.8	1.0	666	47,000	7.06
42	1.0	0.9	128	1120	7.83
48	0.75	0.8	200	165	0.696

Source: EPA Report 12060 EGU, March 1971

With the exception of plants 1 and 48, the calculated losses exceeded the reported loss, with this difference ranging from 1.6 to 20 times the reported loss. Because of the compensating errors in the plant reported losses, such as sampling and testing errors, under-filling packages, etc., the reported losses do not accurately reflect the actual milk solid loss to sewer for most dairy plant operations.

The high values back-calculated from BOD appear to be real. Thus evidence is presented of the potential profitability of waste control and the value of accurate BOD and compositional analysis of dairy food plant wastewaters as a quality control tool in reducing fat and solids losses in plant operations.

CONTROL OF WATER USE

Minimization of wastewater volume and strength is necessary to reduce the waste coefficients found in most plants to the achievable level of about 0.5 pound wastewater/lb. of milk processed and 0.5 pound of BOD_5/1,000 lbs. of milk processed. This can be achieved with current technology, if management makes the necessary effort. To optimize the control of waste in the dairy industry, the following suggestions are made for plant operations.

1. Segregation of all major sources of waste with separated drains as required being put into all new plants and remodeled plants.

2. Segregation or recycling of nonpolluted water.

3. Utilization of the automation system of the plant to eliminate waste; collecting the first rinses from tanks and cleaning operations, and saving the water-milk mixtures resulting from startup and changeover shutdown of HTST units.

4. Eliminating all product discharged during startup from product changeover of HTST pasteurizers.

5. Improve the efficiency of CIP and sanitizing operations to reduce required concentration on these materials.

6. Use of post-cleaning rinses as make-up water for sanitizing and/or cleaning.

7. Elimination of returns from wastewater streams.

8. Collection of lubricants and reclamation for reuse.

Implementation of these practices can reduce BOD_5 coefficients to about

1.0 pound in a well managed plant. Water and waste management
implies the supervision and regulation of water and wastes with executive
ability and control (26). The implications of a water and waste control
program are that one should:

1. Expect plants to reduce their water usage and water
   bills by 50%.

2. Expect plants to reduce their wasteloads or surcharge
   bills by 60%.

3. Expect plants to become better friends with their cities,
   maybe even get a public relations boost.

4. Expect employees to be more knowledgeable about plant
   operation as result of greater emphasis on water conservation and waste reduction.

5. Expect the plant profits to rise.

6. Expect that more plant specialization will occur, since
   satisfactory methods of processing and waste disposal
   can only be developed for limited product mix and
   continuous operations.

The need for education in the water and waste area is evidenced by a test
given to Dairy Short Course students at N.C. State University (14). Several simple questions were asked including:

1. How much water does your plant use/day?

2. How much does water cost/1,000 gallons in your town?

3. What is a sewer charge?

4. What is a surcharge?

5. What is pollution?

Water use answers varied from unknown to 500 gal./day to 1,000,000
gal./day with only approximately 20% of the students being even close to
their plant figures. Water cost/1,000 gallons answers ranged from $70.00
to $0.18 with most not answering the question. Several answers indicated
a vague familiarity with the fact that water is sold by cubic feet. Many
had a reasonable answer for sewer charge but the surcharge completely con-
fused them and less than half answered the question.

Half of the respondents said a surcharge was a tax. Pollution became defined in a vague sense. The answers pointed out that most of these dairy
plant employees did not have any idea about these terms essential to water

and wastewater management. Further evidence is given by a recent national survey which pointed out that the dairy industry has developed a
passive attitude toward the control of wastes (2). Only 10.5% of the 624
companies interviewed had information about their Biochemical Oxygen
Demand (BOD). Every plant that had information was (1) treating their
waste, (2) paying a surcharge, or (3) had been cited by a regulatory agency
for noncompliance. Although the survey only covered 10% of the dairy
plants in the country, the plants surveyed process over 50% of all the milk
processed in the United States. Most (87.1%) of the plants discharge their
waste to a municipal system.

A plan for water and waste management for the dairy industry has been developed. The basic premise of the plan is that water should be managed as
a raw product with a real cost. Also, that wastes going into the wastewater
are product losses which cost dollars and that surcharges are going to be
added to further tax these lost dollars. An outline of a proposed program
follows:

Objectives

1. Acquaint dairy plant personnel with water and waste
   terminology.

2. Relate dairy waste losses to wastewater characteristics
   and "our" environment.

3. Relate dairy waste to financial losses in dairy plants.

4. Relate past surveys of individual plants and compare
   them with state averages.

5. Develop and initiate action programs to reduce wastes
   within the plant.

The Program

A. Management Phase — Two hours — involves waste
   terminology, research and survey findings, costs and
   product losses, surcharges and introduction and des-
   cription of water-waste supervisor and employee phases.

B. Water-Waste Supervisor Phase — Two hours — instruct
   water-waste supervisor (appointed by management) in
   respect to recent research findings; plant activities to
   be carried out in the "employee phase" of the program.

C. Employee Phase (may involve all or selected employees) —
   "Washing Profits (Your Salary) Down the Drain" — four
   hours — illustrate good and bad waste practices; explain
   good waste and water management; relate product losses

to cost and familiarize the employees with waste
terminology and current techniques in controlling
water and waste in the dairy plant. Stimulate and
involve the employees to suggest solutions within
their area of responsibility in the overall attack on
the plant water-waste problem.

D. Follow-up Phase — Two hours — A tour and evaluation
session conducted with management and the waste supervisor to determine progress made within the plant.

Since the water-waste supervisor will be management's representative responsible for the total water-waste program, his responsibilities are outlined
as follows:

1. Every dairy processing plant should have one person
   who is completely responsible for the water and waste
   program. He should have latitude to develop the program and the authority to implement his plans reporting directly to the general manager. He should not
   be overloaded with duties which would keep him from
   devoting at least a majority of his time to water and
   waste problems.

2. A survey of the plant must be made. Sketches, draw-
   ings, and maps should be detailed that indicate the
   size, capacity, and location of water lines, meters,
   sewer lines, junctions, manholes, and other parts of
   the system. Operating data should be compiled relating
   to water used, waste generated, as well as the BOD and
   flow from the main sewer. This data should indicate the
   production rate per hour, per operating shift or per batch
   and it should be readily available for management studies.

3. The water-waste supervisor and a management team should
   then examine the plant critically. Examine why you are
   using water and where the wastes are coming from. Plan
   to reduce water and wastes.

4. Using the results of your management team, attack the
   major water users or waste contributors. Install nozzles
   on hoses. Use level controls to prevent losses. Replace
   leaking water or product valves. Check your machine
   maintenance program; is it up-to-date and sufficient to
   prevent losses?

5. Simultaneously, a water and waste savings educational
   program should be instituted. Employees must be informed

and involved in the program. The supervisor must be responsible for the program, but it is the operating personnel who will determine if your program will be a success.

6. If pretreatment or treatment is required, engage an engineer to recommend solutions.

7. Remember the program must be continuing or it will die. Mention of the program must be frequent. The supervisor must follow the operations assuring that proper water and waste procedures are followed.

The following practices have been recommended to help dairy processing plant managements control their water and wastes (2)(4)(14):

1. Instruct personnel in the proper operation and handling of equipment.

2. Mark all valves clearly, especially multiport, so that it is practically impossible for inexperienced help to turn the valve the wrong way.

3. Handle with extreme care all sanitary fittings, valves, rotary seals, pump parts and filler parts during every phase of operation to prevent marring which may cause leaks.

4. Provide accurate temperature controls on plate, tubular or surface coolers to prevent freezing-on, especially with glycol systems.

5. Install suitable liquid level controls with automatic pump stops, or other devices at all points where overflows are likely to occur.

6. Never fill cheese vats, ice cream mix vats, or cooling tanks to such a high level that spillage will occur when the product is agitated.

7. Worn-out or obsolete equipment or parts of equipment, including sanitary valves, fittings and pumps should be repaired or replaced. When a leak cannot be repaired during processing, buckets should be provided so that the product is not allowed to go down the floor drain.

8. Foam contains a considerable amount of milk solids and should be kept out of the sewer. Common sources of excessive foaming are open-type separators, splashing when filling tanks, air sucked in through leaky

connections in lines under partial vacuum, through leaky packing and through faulty rotary seals or pumps.

9. When unloading tank trucks, avoid spillage from the sanitary pipe or hose connections. Be certain the tank is empty before disconnecting the sanitary pipe or hose.

10. Do not use a constantly running water hose in any room. Eliminate the cause of spillage, rather than just wash it away after it has occurred.

11. Where sugar is used in the manufacture of dairy products, take care that it is not spilled on the floor and washed into the sewer. Sugar will pollute the stream just as much as will milk solids.

12. Provide a foolproof whey collection system and avoid leaks from valves and fittings on whey lines which will result in ultimate loss of whey to the sewer and in floor corrosion.

13. Provide standby pumps for pumping whey from cheese vats to storage tanks in case of power or pump failure.

14. Provide whey storage tanks of twice the maximum daily volume for all types of cheese and casein plants.

15. Sweep up all spilled cheese curd particles from the floor. Do not wash them down the sewers.

16. Pipelines should be installed so that they are properly supported to eliminate vibration-induced leaking joints (where welded lines are not used or points in welded lines where CIP gaskets are utilized) and to insure that the lines are properly pitched to insure draining of the lines.

17. All cases, conveyors, and stackers should be maintained in proper operating condition to avoid jamming and subsequent loss of product from spillage and/or broken packages.

18. Water hoses should be turned off when not in use. Hoses equipped with automatic shut-off valves should be utilized to avoid excessive water usage.

19. For plastic and glass bottle fillers, cappers should be maintained in first class condition to avoid breakage and product loss. Similarly, paper filling machines should be

    maintained in order to avoid jams and product
spillage. Fillers should be equipped with drip
shields and the spills collected to avoid product
going to drain.

20. Filler valves should be checked to see that all con-
taliners are filled to correct capacity at the temperature filling. In glass bottles, filling to the cap seat
may create spillage when the milk warms up by the
cap being forced past the cap seat.

21. Bottles and paper containers should be handled carefully during casing, stacking, loading and delivery
to avoid product losses to drain.

22. Where sweetened, whole or condensed skim milk or
ice cream mix is manufactured, care needs to be taken
so that sugar or dry ingredients are not spilled on the
floor and washed into the sewer. Any spillage of dry
ingredients should be swept up rather than washed down.

23. Care needs to be utilized to avoid incorporation of
cleaning compounds and/or sanitizing solutions into
milk products and thus creating excessive materials
for disposal.

24. Adequate weirs and continuous sampling equipment
should be provided at the outlet of all dairy plants
for monitoring wastes, strengths and volume.

25. Eliminate steam-water mixing T's for making hot water
and utilize regular hot water systems to permit use of
shut-off valves at the end of hose lines. This can result in a 40 to 60% reduction of wastewater volume.

26. Thoroughly drain all lines, tanks and processing vats
before rinsing. Rinse before product dries on surface.

As part of the study of a typical North Carolina plant, the Pine State Creamery, reported by R.E. Carawan, V.A. Jones and A.P. Hansen (14), water
meter data was used to calculate use for specific locations, operations,
areas or products. Data has been summarized into tables and graphs repre-
senting average and typical results.

Considerable variation in water used per day and in volume of product produced per day suggested that water use be considered in terms other than
daily use. Therefore, water use data is presented either in gallons of water
use/1,000 lbs. of product or as % of total use. The average plant water
use is presented in Table 47.

TABLE 47:   PLANT WATER USE

	Total Products, gal./1,000 lbs.
Processing plant	311.4
Office complex	2.2
Refrigeration shop	1.1
Garage	10.8
Total	325.5

Source: PB 220,704, December 1972

The processing plant was such a large percentage of the total use that primary attention was focused on this area. The total use of 325.5 gallons per 1,000 pounds of total product can be compared to the survey figures presented for the United States dairy industry (2). Converting the wastewater coefficient presented in the survey to gal./1,000 lbs. gives a range of 168 to 468 and an average of 302. These survey values were based upon wastewater discharge whereas the figures for this study are in terms of water use.

The effluent averaged 74% of the water use at Pine State Creamery. Another obstacle in comparing values is the fact that water use was based on total product processed while the survey was based upon milk received. Average values adjusted to correspond to the survey data are 300 gallons of wastewater discharge/1,000 lbs. milk which approximates the survey average of 302 gal./1,000 lbs. milk.

The average water volume used to process various products is summarized in Table 48.

TABLE 48:   AVERAGE WATER VOLUME

| Product | Gallons Water Used/1,000 lbs. Product | |
	Average*	Range
Fluid products	259.6	80.0- 736.2
By-products	419.5	39.9-3,067.0
Frozen products	1,397.0	137.0-4,600.0
Total products	311.4	182.0-1,131.0

*May 18 - July 14

Source: PB 220,704, December 1972

Although there was a wide range of values for each product group, the averages permit comparison of the relative quantities of water used in processing. Frozen products processing required far more water (1,397 gallons per 1,000 pounds) than fluid products or by-products. Forty percent of the water use in the frozen products area was for the Vitaline. Vitaline production was higher than normal during this period and accounts in part for the high water use value.

Comparing the water use for frozen product processing with Harper et al (2) survey data is impossible because their values were based upon milk received whereas the present value is based upon final product which included all ingredients. The 1,397 gal./1,000 lbs. frozen product was somewhat below the ice cream plant average of 1,870 gal./1,000 lbs. product cited by Harper et al.

Although water use per unit of frozen product was high, the total use in this area was dwarfed by the fluid products processing area as shown in Figure 44. This is due to the much larger production of fluid products.

FIGURE 44:   WATER USE FOR PRODUCT GROUPS

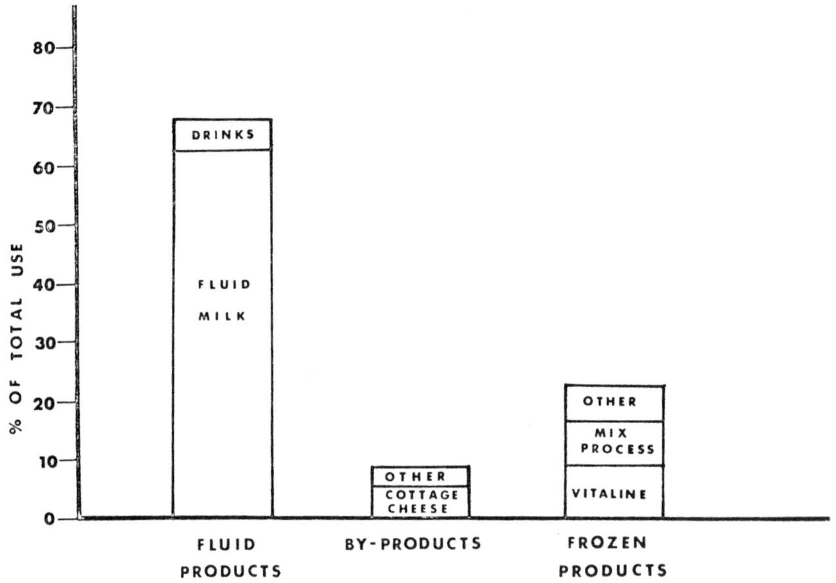

Source: PB 220,704, December 1972

The major water uses in fluid products processing are presented in Figure 45 on the next page. The utilities group required largest input of the water.

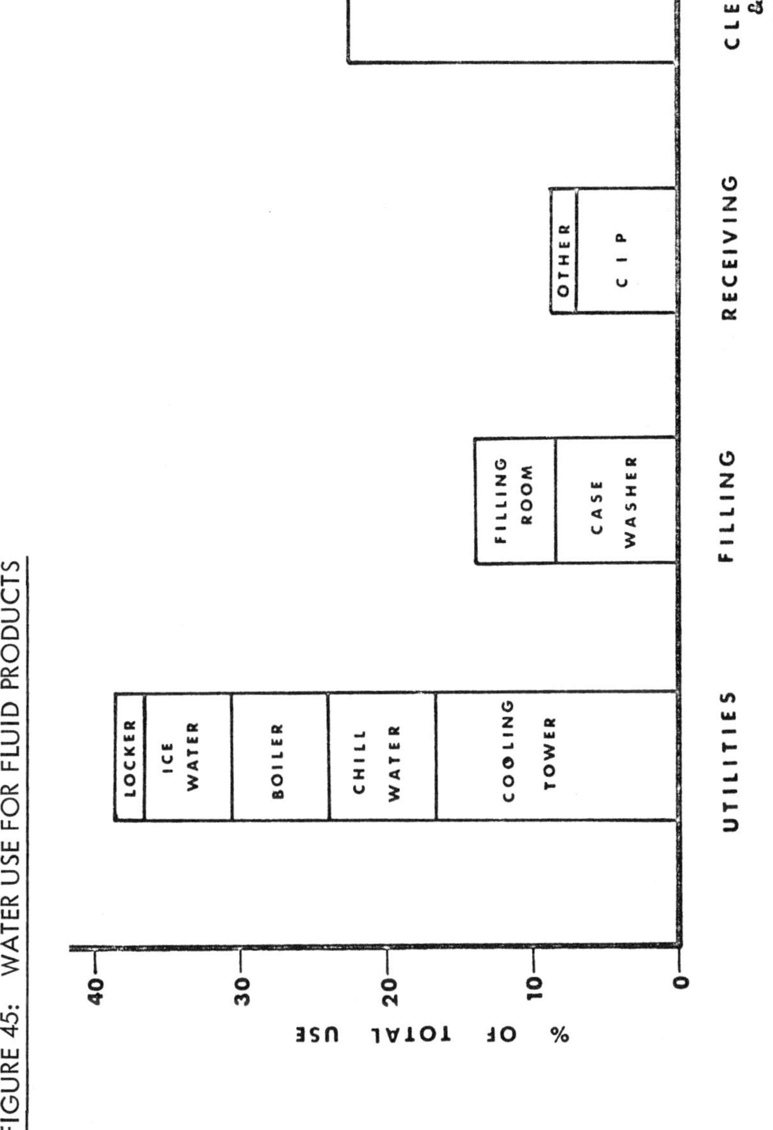
FIGURE 45: WATER USE FOR FLUID PRODUCTS
LOCKER
ICE WATER
BOILER
CHILL WATER
COOLING TOWER
FILLING ROOM
CASE WASHER
OTHER
CIP
CLEAN. & SANIT.
RECEIVING
FILLING
UTILITIES
40
30
20
10
0
% OF TOTAL USE
Source: PB 220,704, December 1972

The breakdown of the various utilities are shown in Figure 46. Cleaning and sanitizing was the next largest area of water use.

FIGURE 46:   WATER USE FOR SELECTED OPERATIONS

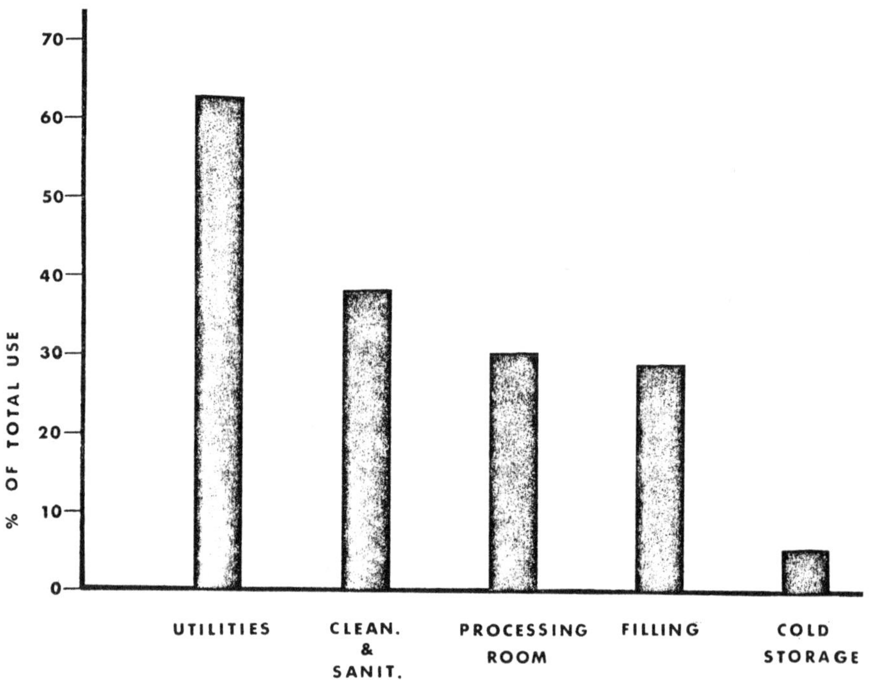

Source: PB 220,704, December 1972

The filling area deserves special consideration. Note that the case washer uses 8.4% of the total water used in the plant. Ways to reduce water use have been mentioned and discussed previously.

Average water use is presented in Table 49 shown on the following page for selected operation and/or locations. The water use for the case washer totals approximately one gallon for each case of milk processed. The utilities group is again prominent as would be expected considering that 38.8% of the total water used in the plant is for utilities.

The consumptive use figure was undoubtedly higher than normal because the data collection period occurred at the peak of drink production.

TABLE 49: WATER USE FOR SELECTED OPERATIONS AND/OR LOCATIONS

Operation or Location	Product	Gallons Water Use/1,000 lbs. Product Average (gal./1,000 lbs. product)*	Range
Receiving	Receipts	33.0	16.9 - 827.9
Total use	Receipts	389.9	264.4 - 1,130.0
Case washer	Fluid products	32.0	2.2 - 89.4
Filling room	Fluid products	22.0	13.2 - 49.6
Cheese room	By-products	255.3	13.2 - 2,503.0
Cold storage	Fluid products and by-products	9.2	3.4 - 93.2
Boiler	Total products	21.2	12.2 - 85.1
Cooling tower	Total products	51.8	23.7 - 215.9
Locker room	Total products	6.0	0.7 - 174.0
Receiving	Total products	26.4	12.7 - 110.5
Utilities	Total products	120.7	67.7 - 526.0
Clean and sanitize	Total products	70.4	38.0 - 210.8
Consumptive	Total products	21.7	2.4 - 53.4
Total products	Total products	311.4	182.0 - 1,131.0
Vitaline	Vitaline (doz.)	2.3 (gal./doz.)	0.4 - 8.2 (gal./doz.)

*May 18 - July 14

Source: PB 220,704, December 1972

The average water use by hour for a one week period is presented in Figure 47 shown on the following page. Extremely low values for nonproduction periods such as Saturday afternoon were eliminated before averaging. Graphically pictured is the water use pattern during an average process day. The low use period was found to be in the early morning hours while the 7 A.M. use was more than 65% above the average use.

The 2 P.M. to 3 P.M. usage was high as expected because the majority of the plant cleanup begins during these hours. The wide variation observed must be considered in all calculations for design of water and sewer systems for dairy plants. Also, the hourly averages surely dampened out extreme periods of water use. High and low values are hidden in the averages.

The bihourly use patterns are presented in Figures 48 and 49. The use for frozen products goes up dramatically in the early afternoon. This was probably due to the cleanup operation which is magnitudinous for the freezing area as many pieces of equipment exist for processing and each must be thoroughly cleaned and sanitized.

Product ingredient water was required in the afternoon after the busy ice cream and ice milk novelty and bulk freezing processes were finished. Water usage for utilities and for cleaning and sanitizing almost parallel each other in their afternoon climb. Again, this is related to the cleaning and sanitizing operations in the plant.

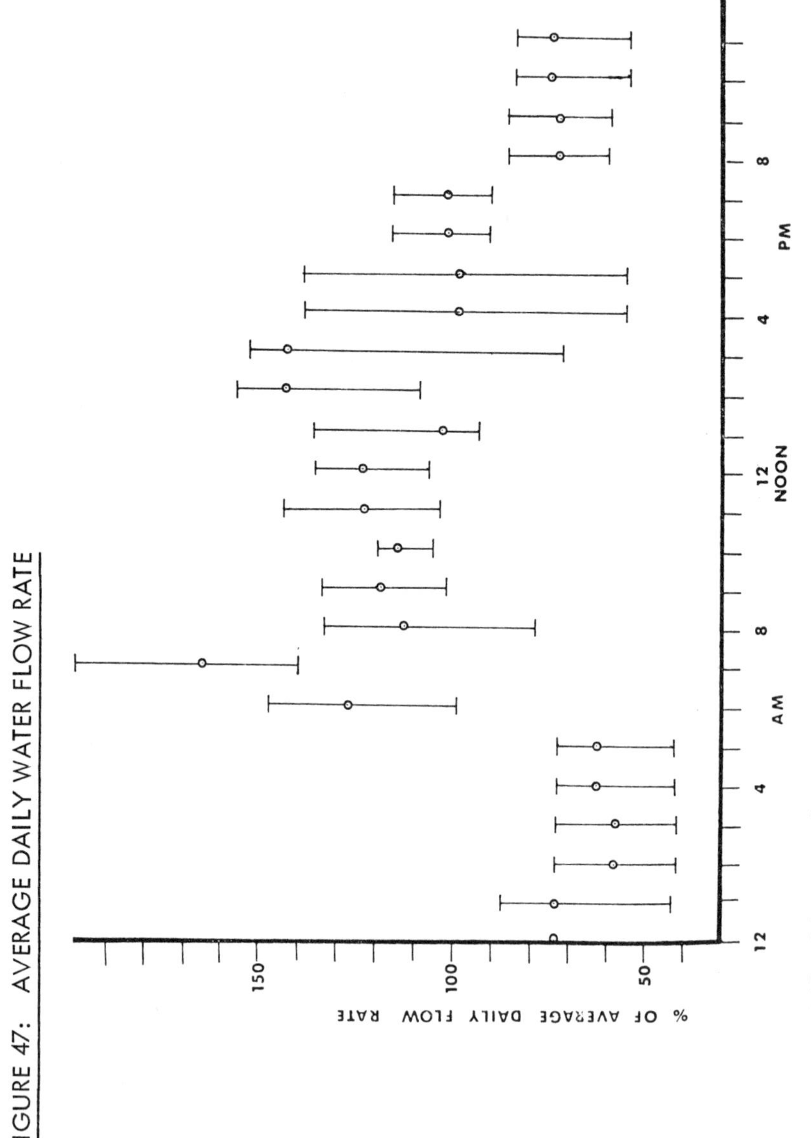
FIGURE 47: AVERAGE DAILY WATER FLOW RATE
% OF AVERAGE DAILY FLOW RATE
150
100
50
12
AM
8
4
12
NOON
4
PM
8
Source: PB 220,704, December 1972

FIGURE 48: WATER USE FOR A TYPICAL PROCESS DAY

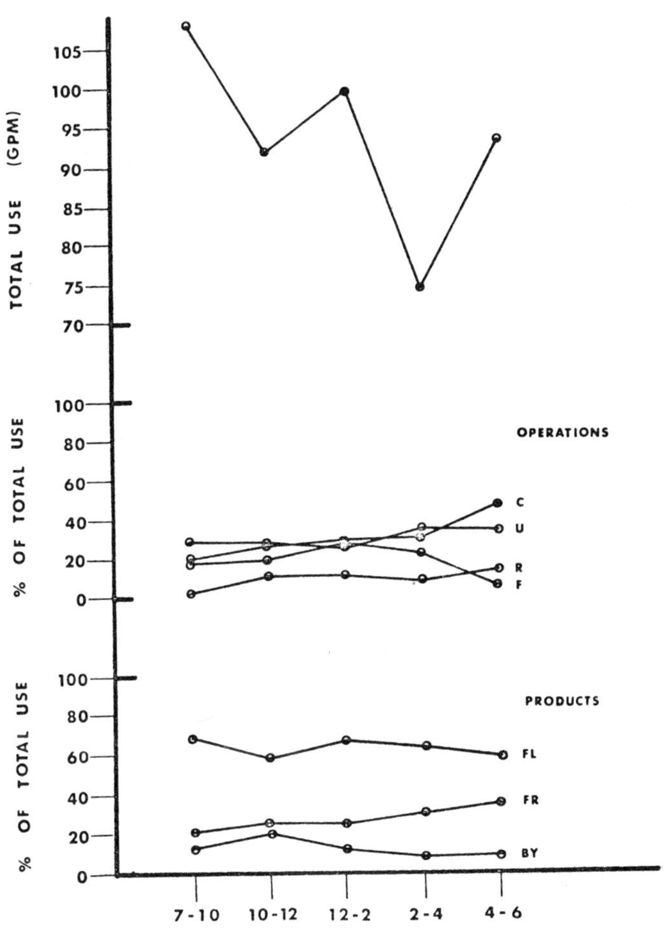

Identification: C – Cleaning and Sanitizing,
U – Utilities, R – Receiving, F – Filling,
FL – Fluid Products, FR – Frozen Products,
and BY – By-Products

Source: PB 220,704, December 1972

FIGURE 49:   WATER USE FOR A SECOND TYPICAL PROCESS DAY

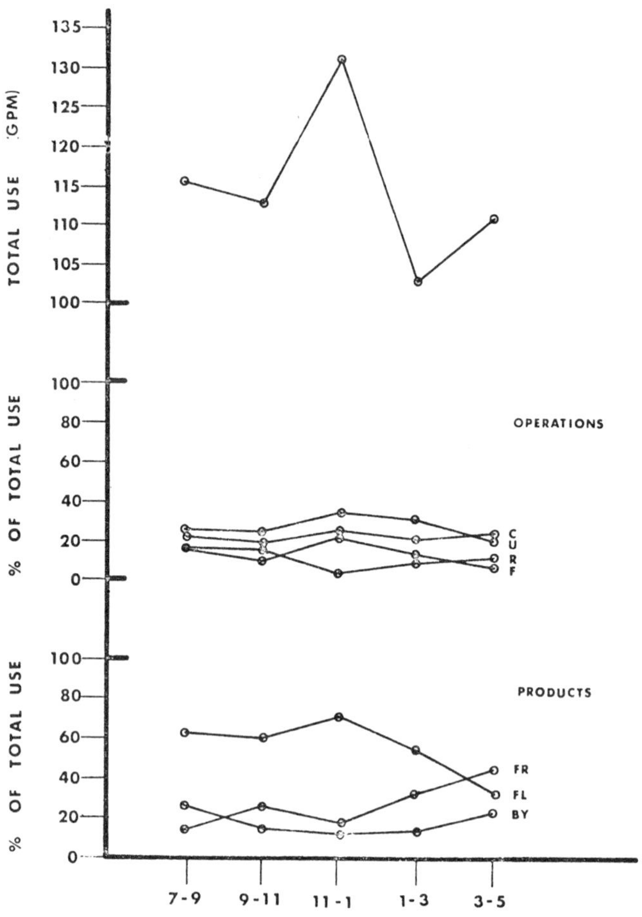

Identification: C – Cleaning and Sanitizing,
U – Utilities, R – Receiving, F – Filling,
FL – Fluid Products, FR – Frozen Products
and BY – By-Products

Source: PB 220,704, December 1972

Role of Condensate in Cooling Waters

Cooling waters not coming in contact with milk are not polluted unless leaks or spillage occur. Condensate waters from evaporating equipment, except for boil-overs, have a very low BOD content and such waters are suitable for direct discharge to stream without treatment or for recycle utilization in other parts of the plant. The BOD strength of such condensates are usually lower than an effluent resulting from the treatment of concentrated milk wastes.

Consideration should be given to the installation of automatic liquid level controls, entrainment separators and other devices in milk evaporating equipment to prevent boil-overs and loss of product, rather than to provide treatment for such wastes. Zack (27) compared the BOD in condensate waters where controls were used and where boil-overs and other losses were known to have occurred. The data are summarized in Table 50.

TABLE 50:   BOD IN CONDENSATE WASTE WATERS

Plant	Parts per Million		Pounds/1,000 Pounds Milk Handled	
	Low Milk Losses	High Milk Losses	Low Milk Losses	High Milk Losses
1	27.0	46.0	0.20	0.51
2	2.5	82.0	0.02	1.48
3	6.0	—	0.15	—

Source: EPA Report 12060 EGU, March 1971

Rinses: An identifiable source of BOD in dairy plant wastes, and one which may be reduced materially, is the product left in lines, tanks, and processing equipment after the product had been removed. The amount of product remaining in a line or processing equipment depends upon the design of the equipment to insure draining, the allowance of adequate time for draining, and the type of product in the equipment.

There is no information available in literature concerning the amount of product remaining on equipment other than milk, so a laboratory experiment was conducted to obtain information on the losses of various fluid dairy products of different viscosities. These values are presented in Table 51 shown on the following page. The relationship of the percentage product remaining to product viscosity is shown in Figure 50.

TABLE 51:   EFFECT OF PRODUCT COMPOSITION ON RESIDUAL
SOLIDS LEFT ON PROCESSING EQUIPMENT SURFACES*

Product	Viscosity, cp	% Remaining on Surface	Pounds BOD_5/1000 Pounds Product
Milk, skim	1.4	0.10	0.072
Milk, whole	2.0	0.26	0.26
Chocolate milk	21.0	0.47	0.69
Half-and-half	15.6	0.39	0.61
Cream, 18% fat	45	0.73	1.5
Cream, 40% fat	91	0.95	3.8
Cultured buttermilk	500	2.3	1.47
Sour cream	9000	3.5	7.0
Ice-cream mix	121	0.86	3.4

*Values obtained following two minutes of drain-
ing. Analyses made at 20°C.

Source: EPA Report 12060 EGU, March 1971

The volume of milk remaining in a 10-gallon can after drainage for various
periods is shown below:

Period of Draining (seconds)	Percent of Product Remaining	Pound BOD/1000 Pound Milk	Accumulated BOD
10	0.20	0.2	0.2
23	0.14	0.14	0.34
30	0.13	0.13	0.47
60	0.09	0.09	0.56
90	0.06	0.06	0.61

The BOD coefficient after 10 seconds draining is about the same as reported
for tank and tank truck cleaning. The effect of the method of feeding 10-
gallon cans to the can washer on product losses are shown in Table 52.

Illustrative data for rinse water volume and fat content of rinses for different types of equipment is shown in Table 53.

FIGURE 50: EFFECT OF VISCOSITY ON PERCENTAGE OF PRODUCT REMAINING ON EQUIPMENT SURFACES AFTER 2 MINUTES DRAINING

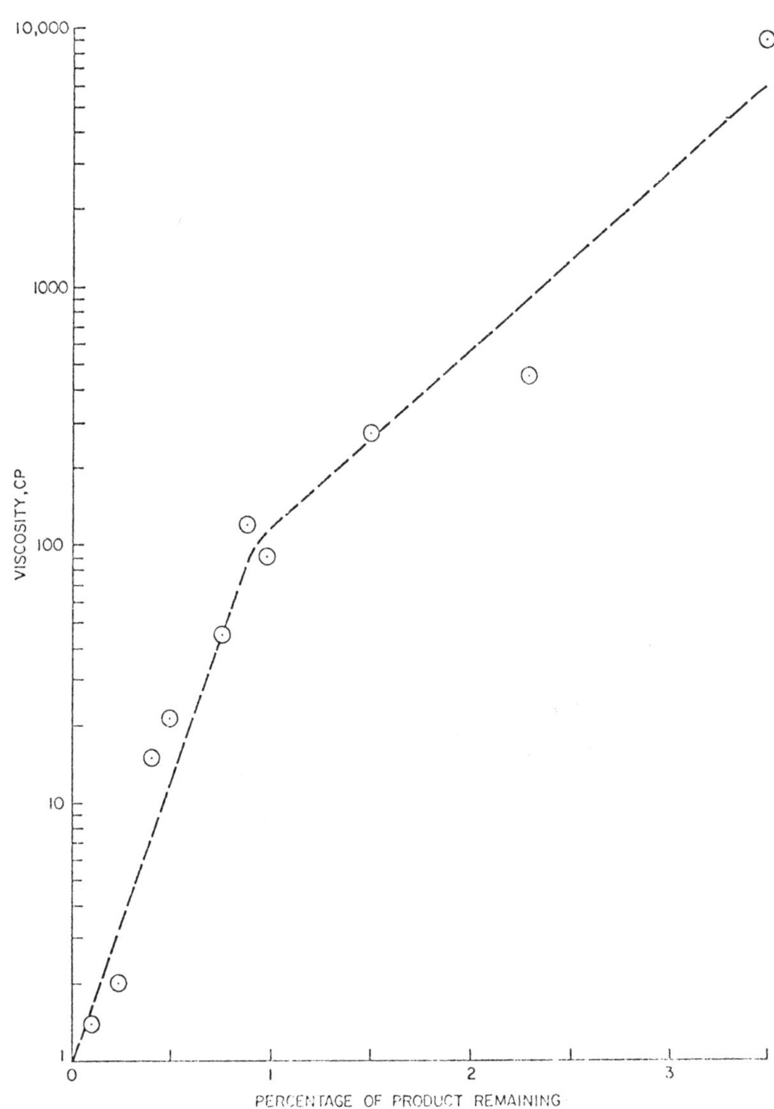

Source: EPA Report 12060 EGU, March 1971

TABLE 52:   EFFECT ON MILK LOSSES OF METHODS OF FEEDING CANS TO WASHER (1,000 10-GALLON CANS)

Disposition of Milk	Fed Improperly		Fed Properly	
	Milk	% of Total	Milk	% of Total
Into weigh can	70,100	99.424	70,617	99.690
Recovered by natural drainage	337	0.478	158	0.223
Unrecovered, left in cans	69	0.098	62	0.087
Total contained in cans	70,506	100.000	70,837	100.000
Milk recovered: per can	0.337	--	0.158	--
BOD$_5$/1000 lb. milk received	0.48	--	0.23	--

Source: EPA Report 12060 EGU, March 1971

TABLE 53:   VOLUME AND FAT CONTENT OF RINSE WATERS FROM VARIOUS TYPES OF DAIRY FOOD PLANT EQUIPMENT

Plant Equipment	Rinse Water, Pound	Butter-fat, Percent	Recoverable Fat, Pound/100 pound
Storage tank -- 3,000 gal.	120	1.20	1.4
Storage tank -- 6,000 gal.	170	2.30	3.9
Homogenizer	62	2.05	1.3
Cooler and connecting pipe	100	1.20	1.2
Product lines	68	2.20	1.5
Cream vat -- 600 gal.	80	2.75	2.2

Source: EPA Report 12060 EGU, March 1971

WASTE REDUCTION THROUGH IMPROVED DESIGN

As new plants are built or remodeled and more and more plants incorporate CIP and process automation capabilities, the proper application of design of plants and process can afford material reduction in wasteloads. The theoretical effect of advanced technology on reduction of wasteloads is well illustrated in Waste Profile No. 9 (3), even though such reduction had not, in fact, occurred in real practice within the industry at the present time. Indeed, in some cases, utilization of automation and mechanization had increased the BOD coefficients. Part of this increase waste relates back to the fact that large complex plants are more difficult to manage and to certain inherent characteristics of mechanization automation which increase wasteloads. Some of these include:

1. Utilization of chain and belt conveyors requires the utilization of relatively high quantities of lubricants which are high BOD-containing materials.

2. Incorporation of casers, case stackers and conveyors increase spillage and loss of product to drain in relatively large volumes because of the dumping in entire cases or case stacks if the equipment is not continuously maintained in good operating condition.

3. Utilization of high capacity equipment up to 100,000 pounds/hour. In the case of pasteurizers, the start-up, shut-down or product changeover of a pasteurizing unit operating 100,000 pounds an hour will cause product loss of about 1,700 pounds (170 lbs. BOD/minute) diverted to drain.

4. CIP separators which shoot sludge to the sewer every 30 minutes have increased the wasteload. With hand-cleaned separators, it was possible to scrape out the

> sludge and handle it as a solid waste.
>
> 5. The product lines in large capacity plants that are CIP
> cleaned hold from 0.8 to 1.2% of the total milk re-
> ceived. Proper design of piping systems, maximum
> utilization of gravity draining, adequate time for
> draining product before initiating the washing cycle,
> and proper designed air blow-down are essential to
> avoid excessive losses.

ENGINEERED REDUCTION OF LOSSES FROM HIGH VOLUME HTST PASTEURIZING UNITS

The industry recognizes that HTST units must be started and stopped on
water. The problem of start-up and changing from water to product, and
back to water, constitutes the major source of losses to drain in the pasteurization process. In starting up the operation of the balance tank, the pasteurizer generally operates at a volume level of about 60 gallons of product, which normally in start-up becomes a mixture of water and product
which must be disposed of, generally down the drain.

Most plants start and stop pasteurizing units 2 and 3 times a day and frequently use water between product change-over. Conservatively figuring
6 such changes a day, approximately 150 lbs. of BOD would be generated.
Through utilization of a recycle system (Figures 51 to 53), material reduction in losses can be achieved. The recycle system is identified at different
stages of product change-over and provides procedure for automatic time
separation of water from product start-up, automatic time separation of
product from water at shut-down and automatic time separation of one product from another product during the operating day.

Through proper design of the balance tank to provide a pod-type outlet and
incorporation of the recycle control system, the volume in the tank at point
of change-over can be reduced from 60 to 15 gallons. The rate of change
in fat content going from water to milk is shown in Figures 54 and 55. Figure 56 shows the start-up and shut-down operation going from water to milk
in a typical system operating at 1,000 lbs./minute (60,000 lbs./hr.). At
the mid-point of a product separation there is approximately 1,000 lbs. of
water-milk mixture at a 1.5% fat test (44 lbs./of BOD). Separation of the
first minute's operation and collection of this in tanks for utilization as a
source of material to be mixed with other solids for utilization in other food
products can materially reduce the wasteload.

In HTST operations, this would result in a 44% reduction of BOD from the
pasteurization process. Assuming the BOD coefficient of pasteurization to
be 0.8, the utilization of an engineered recycle and segregation system

FIGURE 51: PASTEURIZATION RECYCLE SYSTEM — INITIATION OF WATER TO PRODUCT SEPARATION

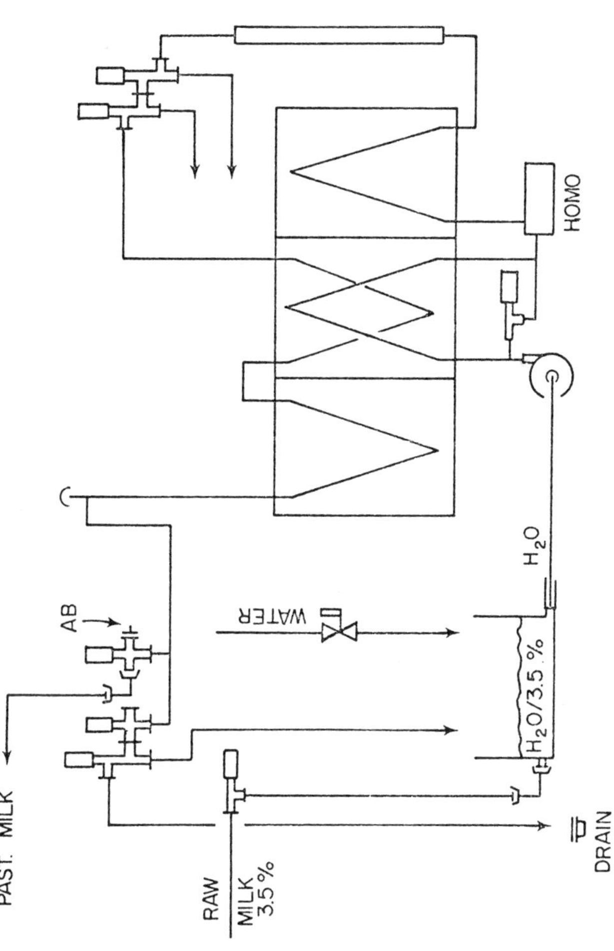

Source: EPA Report 12060 EGU, March 1971

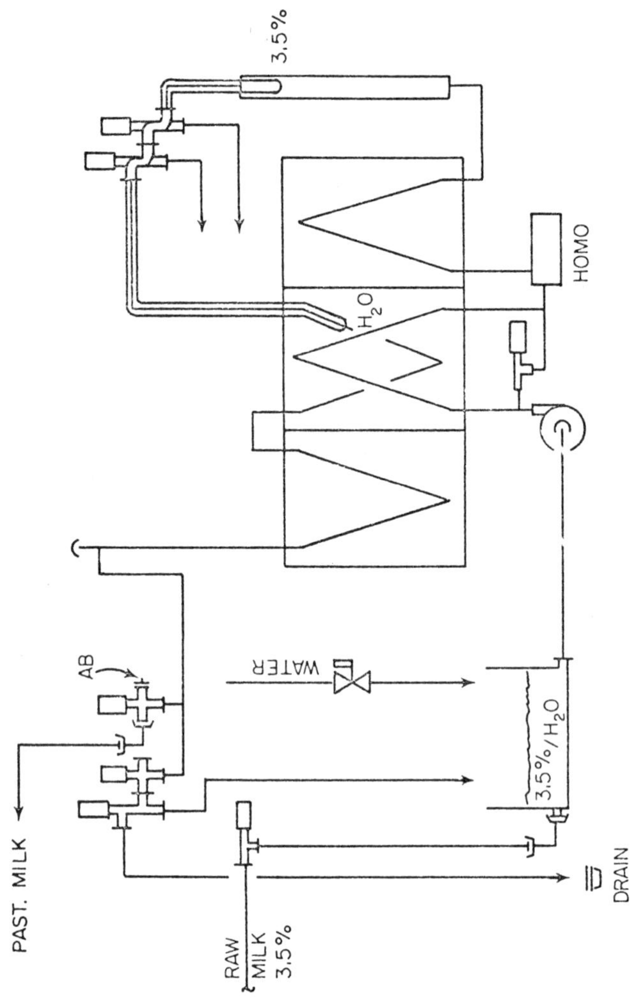

Source: EPA Report 12060 EGU, March 1971

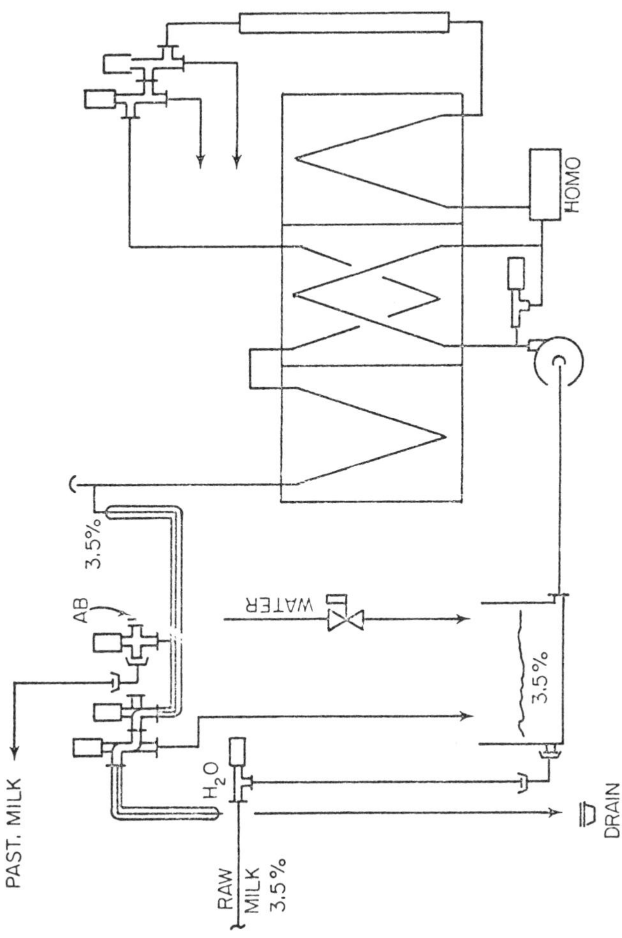

FIGURE 53:   PASTEURIZATION RECYCLE SYSTEM NEARING CHANGE IN RECYCLE VALVE

Source: EPA Report 12060 EGU, March 1971

FIGURE 54: RATE OF CHANGE IN FAT CONTENT

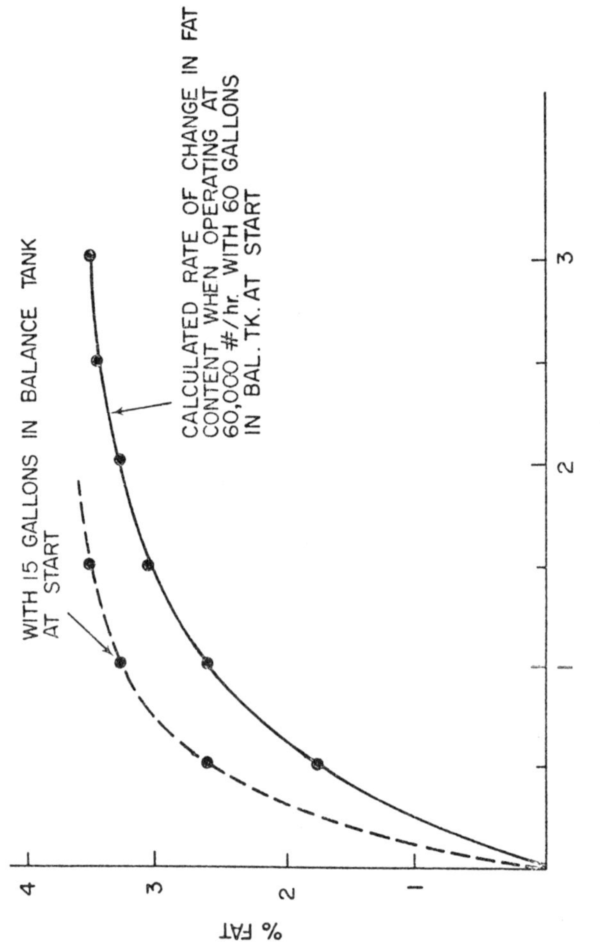

Effect of time after change from water to milk on fat content of wastewater going to drain during pasteurization start-up

Source: EPA Report 12060 EGU, March 1971

FIGURE 55: FAT LOSSES AS A FUNCTION OF TIME

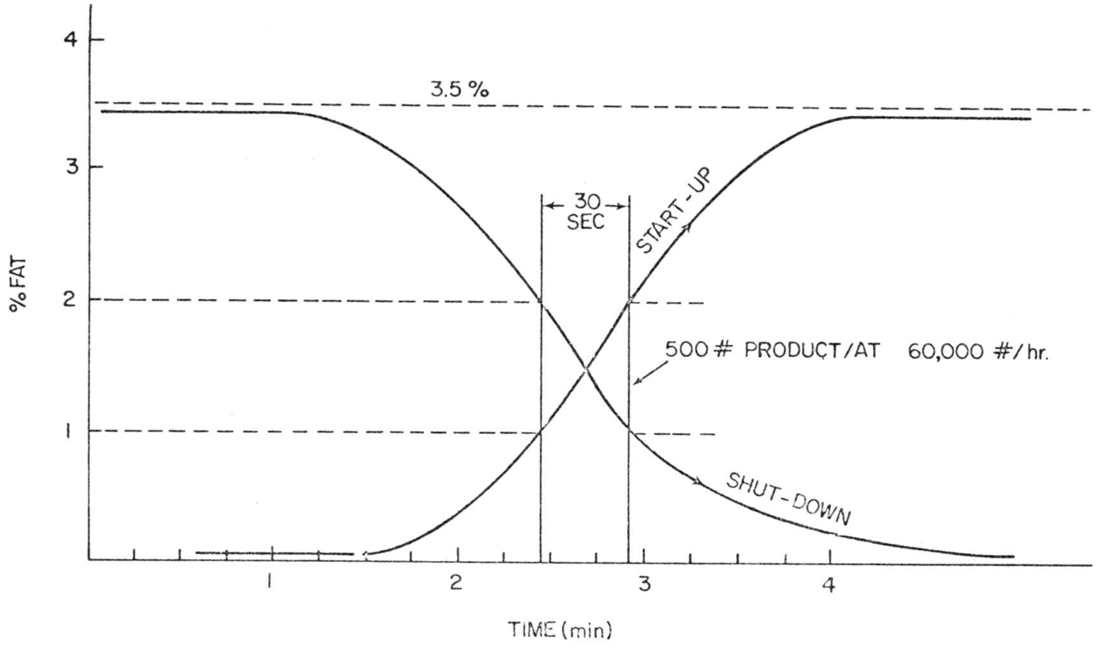

Startup and shutdown of a 60,000 pound/hour HTST pasteurizer

Source: EPA Report 12060 EGU, March 1971

FIGURE 56: DIAGRAMMATIC PRESENTATION OF AUTOMATED SYSTEM FOR MINIMIZING MILK LOSSES

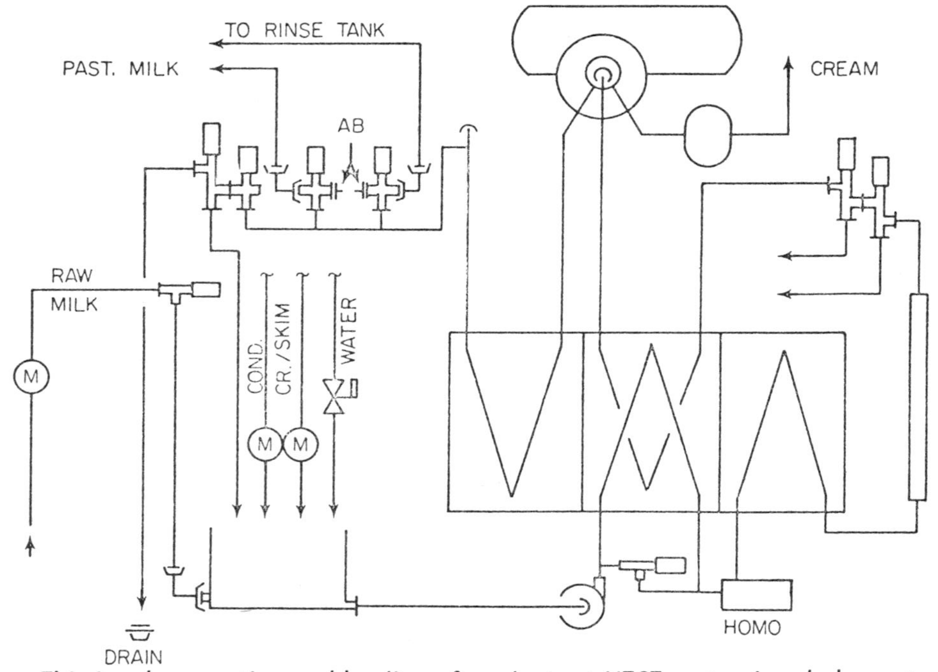

This involves continuous blending of products at HTST pasteurizer balance tank, eliminating need for pasteurizer shutdown during product changeovers

Source: EPA Report 12060 EGU, March 1971

can reduce this coefficient to 0.45, which is about the value reported for
two plants utilizing the recycle system.

Dairy food plant automation systems can be utilized similarly in recovering
rinses from tankers, tank trucks and lines and collecting these for product
use as well. In one operation, the rinsing of 6,000-gallon tank trucks utilized 250 gallons of water for a single rinse, resulting in about 9.0 lbs.
of BOD in 250 gallons of water per tank truck. By changing the three burst-
rinses, segregating and saving the first 30-gallon rinse, 80% of approximately 7 1/2 lbs. of BOD was eliminated from the sewer per tank truck.
This rinse contained 1.5% fat which reduced the receiving coefficient to
0.05. This fat content of rinses raises to 3.4% fat for high solids products
or for rinses from tank trucks that have remained stationary for more than
1 hour before being emptied.

REDUCTION OF LOSS THROUGH CONTINUOUS PROCESSING

The utilization of the centrifugal machine in the form of a clarifier-sepa-
rator can be utilized in combination with high-temperature short-time systems to accomplish a variety of processing operations in a simple and
straightforward manner to eliminate the necessity to utilize processing
vats for the make-up of skim milk, low-fat milk, fortified milk, half-and-
half, creams and other by-products. Utilization of positive displacement
pumps may serve as a means of (a) providing for improved control of fat
test and, (b) insuring the movement of cream to the cooling equipment in
the tanks and vats at various distances without danger of loss of product
through the separator plugging. The combined system could be utilized
on a continuous and automated basis to make a variety of fluid milk products.

Milk can be metered to a constant level tank as may cream, condensed
skim, or plain skim condensed, and in this manner a variety of products
can be produced through "on-the-run" batch blending. By initiating flow
into a given surge tank while still on homogenized milk, and reverting back
homogenized milk for changing to the next surge tank, it is possible to make
one product changeover after another in the pasteurization system with
homogenized as the intermediate, rather than water; thus avoiding the loss
of product resulting from startup and shutdown on water.

At the same time, it will eliminate the utilization of the intermediate tanks
and thus saves the product normally adhering to the walls of these tanks
which would go to the sewer as waste at the rate of about 0.2 lb./1,000 lbs.
of product processed for low viscosity products and up to 3 lbs./1,000 lbs.
for cream.

SEGREGATION OF WASTES

When planning new plants, or remodeling plants that exist, consideration
should be given to the segregation of those sewers receiving high waste in
BOD content from general floor and area drains. One example of new design possibilities would be to combine up-drains at HTST constant level
tanks as CIP units and CIP-type centrifugal machines into a single separate
sewer system. This system may be initially connected to the existing system
outlet of the plant but later, if desired, it could be connected to a retention tank to permit metered discharge of the high wasteload over a prolonged
period of time; it might, in fact, be subjected to some type of pretreatment.
Illustration of such a segregated drain system is shown in Figure 57.

FIGURE 57:   FLOOR DRAIN SYSTEM FOR SEGREGATION OF HIGH AND LOW SOLIDS WASTEWATERS

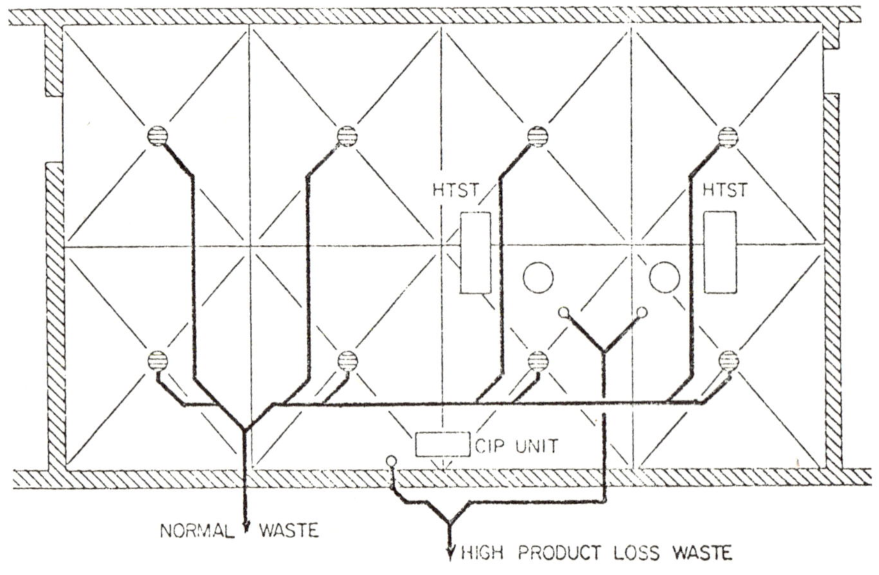

Source: EPA Report 12060 EGU, March 1971

Similarly, lubricants, milk in filling areas, and solid particles in cottage
cheese filling operations might be segregated by a separate drain system
and collected at a central point for pretreatment. The installation of desludging centrifuges on whey operations, especially cottage cheese whey
operations, can materially reduce the BOD level of whey through the removal of fat and curd fines. In pilot trials at The Ohio State University,

the BOD of cottage cheese whey has been reduced from 45,000 to 30,000 by the use of a desludging clarifier and the utilization of the recovered curd as a product for baking purposes.

CONTROL OF WASTEWATER THROUGH BY-PRODUCT AND WASTE PRODUCT UTILIZATION

Because of the national attention and visibility of whey as a waste product, the dairy industry is well aware of the significance of using whey as a food or feed product to minimize pollution and to gain a profit from such operations. They are less aware of the potential in respect to the utilization of product rinses, dilute milk solutions resulting from startup changeover and shutdown of pasteurizers on water, which is a major source of BOD in most food milk plants and the recycling or utilization of returned product.

Whey

The complete removal of whey from wastewaters is a continuing problem in the United States, in part, because of the large volume of materials to be handled (about 20% of all the milk processed ends up as whey of one type or another), the limited market for whey solids, and technological problems in developing products from whey.

A major problem in respect to the elimination of whey from wastewaters lies in the marketing of the product so far as a profitable return on the product is concerned. However, most proposed methods have applicability for converting whey into usable products for feed, if not for food, and can result in a reduction of cost to a dairy operation where they are faced with an absolute limit on BOD in the wastewater or where they are paying a surcharge on BOD at the present time.

In this connection, it might be well to note that one plant recently built to produce food grade whey powder has been diverted to feed grade, a lower profit product, because of the lack of a market for the dried whey which could be attributed, in part, to low-cost imported dry milk solids utilized in preference to whey.

One of the most promising developments in respect to whey utilization lies in the recoverability of the protein fraction, separate from the rest of the whey, which has unique nutritional and functional properties. The nutritional value of the whey protein exceeds that of casein; the protein has unique whipping and emulsifying properties. Removal of the protein from the whey reduces the whey solids about 13% and the BOD about 20%.

Methods for economically removing lactic acid or sodium lactate from acid whey, and practical procedures for converting lactose into its simple components, galactose and glucose, are needed to provide for further market potential of whey solids. At this time, the problem of whey utilization would appear to be more oriented to economical and marketing considerations than to needs for new technology. The available technology would appear to adequately provide for complete utilization of sweet whey as food or feed material. The market and economic possibilities need to be improved.

The problem of whey removal and utilization is such a complex subject that, in addition to the above brief survey paragraphs, an entire chapter will be devoted to the topic in somewhat more detail; see the chapter which follows this one.

Buttermilk Solids

With the increasing utilization of continuous churn processes and with the decline in butter plants, the disposition of buttermilk is becoming a minor problem. Most of the buttermilk in the country at the present time is being withheld from wastewater and is either dried for food or feed use through being sold directly for feed.

Recoverable Milk Solids

As previously indicated, the potential exists for elimination of all the milk solids present in rinse waters from tank truck storage tanks, lines and equipment, saving the milk solids diverted to drain in the startup, changeover, and shutdown of HTST pasteurizers and removal of milk solids put into the wastewater by virtue of returned products. It would appear that up to one pound of BOD/1,000 lbs. of milk processed could be eliminated through the collection and utilization of these solids. In the case of highly viscous products, such as cream, churned buttermilk, sour cream, yogurt, ice cream mix and similar products, the amount of BOD that could be recovered from wastewaters might be as high as 3 lbs./1,000 product in some plant operations.

At the present time, these diluted milk solids are considered to be adulterated products by many health officials and changes in laws and regulations will have to permit their utilization in foods. There is need for further investigation to optimize methods for segregation and utilization of these dilute materials. The possibility exists at the present time of using them in ice cream mix or any other product where solids can be added to the material. The possibility also exists of utilizing osmosis to concentrate the materials. However, this will require additional membrane technology since protein and lipids tend to foul the reverse osmosis membranes currently in use.

To ascertain current opinions in respect to the best methods for dairy waste treatment, newly developed methods and possible problems associated with dairy wastes, inquiries were sent (2) to 175 consulting engineering firms advertising in the Yearbook of Environmental Science as providing consulting services for wastewater treatment. 106 replies were obtained from these inquiries, of which 17 (16%) indicated previous experience with waste treatment facilities for dairy food plants.

The organizations were asked to specifically comment upon their preference for waste treatment methods, their recommendations for the treatment of whey, their familiarity with any new treatment systems particularly applicable to dairy food plant waste handling, specific problems they had encountered in dairy plant waste treatment, and their experiences in respect to the treatment of dairy food plant wastes in comparison to municipal wastes. Personal contact was made with all of the organizations indicating experience with dairy food plant waste treatment. As might be expected, the opinions expressed by the various consulting organizations were varied.

Consulting firms favored the biological oxidation of dairy food plant wastes and indicated the impracticality of other procedures. A common denominator that ran through the comments from all of the organizations was that the best methods of treatment for dairy plant waste would vary from plant to plant depending upon the quantity of wastewater, strength, location of the plant geographically, location in respect to major municipalities, the effluent requirements of streams to which the wastewaters were discharged, land cost, labor market, local construction costs and availability of adequate treatment facilities by the local municipality.

For small volume, and for BOD effluents below 750 ppm, most organizations favored spray irrigation. For larger plants, recirculating trickling filters in conventional or plastic media were favored by 35%, complete mixed activated sludge by 35%, aerated lagoon treatment by 20% and combinations of these procedures by about 10%. A number of the firms indicated that the combination of trickling filter followed by conventional activated sludge has been useful to meet stream requirements of 20 ppm or less in the BOD of the effluent. Several of the engineering firms indicated that for large plants, processing over 200,000 pounds of milk per day, activated sludge systems produced the best effluent.

Several firms indicated that from the standpoint of suspended solids that dairy wastes are easier to treat than municipal wastes. However, it is frequently necessary to provide additional biological treatment to dairy wastes to meet stream standards, which is not necessary for municipal wastes. Several suggested the possibility that detergents could be affecting the efficiency of waste treatment in some biological treatment facilities. Regarding the problems of treating dairy wastes, the following details were cited by a majority of the consulting firms:

     1. Difficulty of handling a high waste strength in con-
       ventional biological treatment facilities especially
       if the pH of the waste has been reduced to less than
       6.0 due to holding.
     2. Problems of shock loading and recommendation for the
       installation of flow stabilization tanks regardless of
       the method of biological treatment.
     3. Problems with settling of sludge in activated sludge
       plants was cited as a universal problem.

Other problems included (a) wide variation of flow, wide variation of concentration of contaminates, (b) additional nutrient requirements for biological treatments, (c) extremely high organic shock loads to existing biological treatment systems, (d) quick development of anaerobic conditions in
any type of equalization basin, (e) odors resulting from the abovementioned
anaerobic conditions, and (f) a low pH of some waste and a high pH of others.

One consultant stated, "Milk or dairy product spills which occur frequently,
can upset the biological treatment system for weeks to months." Other comments included, "Operation and maintenance of independent dairy waste
treatment plants are serious problems." "Detergents used in cleaning dairy
plant equipment can upset the process if they are discharged in slugs as is
frequently the case. This becomes more serious as water volume control im-
proves." "Aerated holding tanks to blend and more uniformly discharge
the waste of treatment facilities are highly desirable, as well as highly
shock resistant biological treatment processes." "It is desirable to monitor
pH in the flow of the waste treatment plant and to neutralize extremely low
pH values by lime or ammonia addition."

All organizations are in agreement in their opinion that whey and the wash
water from cottage cheese operations should be excluded from biological
treatment facilities.

WHEY REMOVAL AND UTILIZATION

Whey is defined in Webster's New International Dictionary as "the serum or watery part of milk separated from the more thick or coagulable part or curd, especially in the process of making cheese."

The manufacture of cheese from either whole or skim milk produces, in addition to the cheese itself, a greenish-yellow fluid known as whey. Whole milk is used to produce natural and processed cheeses such as cheddar, and the resulting fluid by-product is called sweet whey, with a pH in the range of 5 to 7. Skim milk is the starting material for cottage cheese and gives a fluid by-product called acid whey, with a pH in the range of 4 to 5. The lower pH is a result of the acid developed during or employed for coagulation.

Each pound of cheese produced results in 5 to 10 pounds of raw fluid whey. The high organic content of whey leads to a severe disposal problem. However, these organic materials have a high nutrient content and, if properly recovered, could provide useful products. Table 54 shows typical compositions of whey and dried whey solids. Over 70% of the nutrients from skim milk show up in acid whey, including soluble protein and lactose.

TABLE 54: AVERAGE COMPOSITION OF WHEY PRODUCTS

Product	Water	Nitrogenous Matter	Fat	Lactose	Acid	Ash
Raw Cheese Whey	93-94%	0.7-0.9%	0.05-0.6%	4.5-5.0%	0.2-0.6%	0.5-0.6%
Dried Whey	2-6	12-14	0.3-5.0	65-70	2-8	8-12

Source: EPA Report 12060, March 1971

The organic nutrients of whey, which go unused, place a costly burden on
sewage systems and waterways. The biological oxygen demand (BOD) of
whey has been noted as ranging from 32,000 to 60,000 ppm. Most of this
BOD is due to the lactose. Specific BOD values for cottage cheese wheys
are between 30,000 and 45,000 milligrams per liter, depending primarily
on the specific cheese making process used.

Every 1,000 gallons per day of raw whey discharged into a sewage treatment plant can impose a load equal to that from 1,800 people. This is
partially passed into streams in most cases because BOD removal is not
complete. Every 1,000 gallons of raw whey discharged into a stream
requires for its oxidation, the dissolved oxygen in over 4,500,000 gallons
of unpolluted water.

Recent statistics give BODs of about 0.2 pound per pound of cottage cheese
curd. Combining this with production statistics, a total BOD removal of
over 200 million pounds would have been required in 1970 for complete
waste treatment. Considering that at the very best only half of the cottage
cheese whey produced is currently put to good purpose and that curd washwater can contain up to 3% solids, the disposal problem is severe.

WHEY PRODUCTION

As pointed out by F. Groves (29-A), whey is becoming an increasingly
important stepchild of the dairy industry. As a by-product of the cheese
industry, its greater production, along with the increased production of
cheese, is viewed with interest by different groups of people.

Environmentalists may look at increased whey production with some alarm
as a potential pollutant of our rivers and streams. Humanitarians are concerned that such a large percentage of a potential source of excellent protein is wasted. Researchers in universities and industry are looking for new
uses for whey and whey derivatives. Some cheese manufacturers, some whey
processors, and municipalities are looking for the least cost or minimum loss
method of disposal. And everybody is looking for a way to turn a product
that has been an economic liability for years into profitable end uses.

The dimensions of the problem are indicated by a production of 23 billion
pounds of liquid whey per year, equivalent on a BOD basis to pollution by
16 million people. To build sewage plants to accomodate this waste would
cost about $800 million, with an annual operating expenditure of about
$30 million.

In 1970, Wisconsin produced over 600 million pounds of American cheese,
about 46% of the nation's total. Total cheese produced in the state, ex-

cluding cottage cheese, was over 900 million pounds (Table 55). Since 1960, cheese production has increased by more than one third, and dried whey production by almost four times.

In 1962, Wisconsin produced almost 7 billion pounds of liquid whey. About 27% of this whey was processed into dried whey powder. Small amounts of the remaining whey were used for the production of albumin and other whey derivatives. Lactose was frequently produced in conjunction with dried whey powder.

Since Wisconsin is a major dairy state and since data are available on the industry, its primary products and its waste products, it will be considered in some detail here as indicative of U.S. industry.

TABLE 55:   CHEESE, WHEY, AND DRIED WHEY PRODUCTION, WIS-
CONSIN, SELECTED YEARS, 1955 TO 1971 [a]

Year	Total Cheese (thous. of pounds)	% Increase from 1955	Liquid Whey Produced [b/c] (thous. of pounds)	Liquid Whey Per Factory (thous. of pounds)	% Increase from 1955	Dried Whey Production (thous. of pounds)
1955	598,112	100	5,204,000	4,922	100	69,867
1960	641,499	107	5,581,000	6,994	142	87,495
1965	770,398	129	6,702,000	10,589	215	158,802
1967	828,911	139	7,212,000	12,455	253	222,153
1968	847,007	142	7,369,000	13,595	276	240,577
1969	856,593	145	7,539,000	14,871	302	244,202
1970	947,591	158	8,244,000	17,140	348	292,981
1971	986,369	165	8,581,000	18,736	380	332,900

[a] Source: Wisconsin Agricultural Statistics, various issues.
[b] Estimated at 8.7 pounds of whey per pound of cheese.
[c] Does not include cottage cheese curd or creamed.

Source: F. Groves, Bulletin No. 48, University of Wisconsin (July 1972)

The proportion of non-cottage-cheese whey that is dried has been increasing. In 1955, it was about 27%, and in 1970, about 71% (Table 56). While the amount of whey being dried has been increasing, so has the amount of whey being used directly in blends of different products. Many products are blended before drying, and the raw whey used in these items is not reported as being processed. Consequently, any current reports of processed whey underestimates processing of these blends.

TABLE 56:   PROPORTION OF WHEY DRIED, SELECTED YEARS, WIS-
CONSIN [a]

Year	Liquid Whey [b][c]	Dried Whey	
	--------------- (000) ---------------		
	pounds	pounds	percent [d]
1955	5,204,000	69,867	27
1960	5,581,000	87,495	31
1965	6,702,000	158,802	47
1970	8,244,000	292,981	71
1971			78[e]

a/ Source: Wisconsin Agricultural Statistics, various issues.
b/ 8.7 Pounds of whey per pound of cheese.
c/ Excludes cottage cheese whey.
d/ Estimated as in footnote c/9 Table 58 (page 159).
e/ From Table 58.

Source: F. Groves, Bulletin No. 48, University of Wisconsin (July 1972)

The number of cheese factories has been declining in Wisconsin for many
years. In 1950, there were over 1,200. In 1971, there were less than
500 (Table 57). On a percentage basis, the decline has been almost two
thirds. Since the remaining plants are larger than those going out of business, the amount of whey per factory has also been increasing. From 1955
to 1970, the amount of whey per factory more than quadrupled (Table 55).

TABLE 57:   NUMBER OF WISCONSIN DAIRY PLANTS MAKING CHEESE,
SELECTED YEARS

Year	Number	% of 1950
1950	1,279	100
1955	1,057	83
1960	798	62
1965	633	50
1967	579	45
1968	542	42
1969	507	40
1970	481	38
1971	458	36

Source: Wisconsin Dairy Facts 1971, Wisconsin Statistical Reporting Service

This greater concentration of whey among fewer factories can have advantages and disadvantages. If satisfactory methods of whey disposal are not available, larger volumes can compound an already sticky problem. On the other hand, if satisfactory alternatives are available, additional supplies of whey can mean greater economic returns from whey, or give the necessary volume base to make economically feasible processing techniques not available on a smaller scale. In other words, if disposal is a problem, additional volume can make it worse or additional volume can increase returns if disposal operations are profitable.

The small cheese factory is still in a disadvantageous position in whey disposal. Its volume is too small to justify the installation of disposal or processing equipment. It is often uneconomic to install the kind of equipment necessary to insure that the whey is cooled and handled in such a way that it can later be used for human consumption. One alternative is to join with other factories and set up some type of processing or disposal system. Those continuing to dump or field spread whey are frequently subject to strict regulations.

A number of small plants had set up roller drying facilities in the early 1960's. Some of these have now gone out of existence, and roller drying is less important. The trend to more stringent sanitary conditions, and use of more whey in human foods, have contributed to the decline in roller drying.

Plants that have no other viable economic alternative are still paying to have whey hauled from the factory. In 1962, over 150 plants were paying processors to dispose of their whey.

As pointed out by S. Boxer of Dairy Research & Development Corp. of New York City in Proceedings of the Second National Symposium on Food Processing Wastes, EPA Publication 12060, March 1971, Washington, D.C., U.S. Government Printing Office, the economics of the disposition of whey clearly indicate that large-scale processing is necessary. Of approximately 700 cheese plants in the country, only 10% produce sufficient volume to justify their own whey drying plant.

Seasonality of Whey Production

Much of the cheese produced is made from manufacturing grade milk. The seasonal swing of manufacturing milk production is usually wider than the seasonal swing for Grade A milk production. Cheese and whey production follow the same seasonal pattern, which means that processing plants must have the capacity to handle flush volumes or use other disposal methods in flush periods. In 1961, a whey processing plant needed 170% of the November capacity to handle the volume in June. An average of 1969 to

1971 showed that the seasonal swing in production had declined to about
160% of the high month over the low month.

WHEY DISPOSAL

As pointed out by F.W. Groves and T.F. Graf (29), cheese factories dispose of whey by many different methods. The most common are: returning
whey free to farmers, dumping whey as waste or sewage, selling whey to
processors, and paying processors for disposition. However, only a small
proportion is sold.

Cheese factories located in areas of high hog production return most of
their whey to producer patrons, and sell very little to processing facilities.
Factories returning whey to farmers either deliver the whey on their milk
procurement routes, or farmers pick the whey up at the plant. Many factories use both methods of distribution.

Approximately 15% of the plants surveyed by Groves and Graf stated that
they incurred no additional cost in delivering whey free to farmers over
and above their regular milk procurement costs. Others indicated the
necessity of maintaining a tank on a van to haul whey, and the extra labor
necessary to load and unload this tank resulted in additional direct costs.

Over 25% of the factories reported paying processors to dispose of whey for
them. The majority of these (93%) reported a cost of two cents per cwt.
for this type of disposition. Only about 2% of the participating cheese
factories reported that they were able to sell whey to farmers. Another
2% were able to sell their whey to processing plants, and 4% processed
whey in their own plants.

Over one half of the plants reported they disposed of all, or part of, their
whey as waste or sewage. The weighted average estimated cost for this type
of disposal was slightly over one cent per cwt. Over one half of the plants
also indicated they had no other way to dispose of whey other than the
method they were presently using. Some plants said they had alternatives,
such as building their own processing or disposal facilities or using whey
for fertilizer.

As noted above, the most common methods of whey disposal, in 1962, were
returning whey free to farmers, dumping as waste or sewage, selling whey
to processors, and paying processors for collection. The proportion of whey
that was sold was rather small and virtually all was used in processing products for human consumption. All of the above methods of disposal were still
being used in 1972. However, more whey is being processed, and whey
disposed of as waste or sewage is less and is more closely regulated than in

1962. As noted, in 1962, less than 30% of the whey could be accounted
for as being processed. In 1971, less than 10% of all the non-cottage-
cheese whey was unaccounted for, and about 25% of all whey was unaccounted for (Table 58). In 1971, 91% of the non-cottage-cheese whey
was either dried or condensed. However, this estimate of dried whey may
be high because undoubtedly some of the condensed whey later ends up as
dried whey and may be double counted.

TABLE 58:   UTILIZATION OF WHEY IN WISCONSIN, 1971

	(000)	% Without Cottage Cheese	% With Cottage Cheese
Cheese Production (Excluding Cottage Cheese)	986,369		
Pounds of Liquid Whey[b]	8,581,410	-	
Liquid Whey Dried[c]	6,660,000	78	64
Liquid Whey Condensed[d]	1,146,000	13	11
Unaccounted	775,410	9	
Cottage Cheese Whey[e]	1,800,000		
Total Unaccounted	2,575,410		25
Total of All Whey	10,381,410	100	100

Source: Wisconsin Crop Reporting Service.

[b] Estimated at 8.7 pounds of whey per pound of cheese.

[c] Estimated from 333 M P dried whey at 20 # liquid whey per pound of
dried whey.

[d] 30 percent Solids, 573,000 # Solids x 20 # liquid, per pound of solids.

[e] Based on 40 M P Cottage Cheese Production.

Change in Disposal Methods

The change in whey disposal and utilization over the last several years has
been the result of several factors. One is that less whey is being returned
to farmers because of the shift to bulk farm pickup and the high cost of
operating return routes. In 1962, those plants that estimated a cost of returning whey to farmers felt it was in the range of one to ten cents per cwt.
The weighted average estimate of cost per cwt. was 2.79 cents. Many

factories did not attempt to estimate the cost of returning whey, but knew that it existed because of the labor involved and the cost of maintaining equipment.

Other factors include greater volumes of whey to be handled, increased costs of labor and equipment, and more specialization of agriculture. These have all contributed to less whey being returned to the farm. However, whey is still considered an excellent feed, and the University of Wisconsin is currently conducting a research program on using whey in animal feeds.

Increased concern with the environment has led to increased enforcement of whey disposal regulations. Less whey is being indiscriminately disposed of, and more is going into municipal and other disposal systems, or some kind of processing.

Field Disposal

Some whey was being disposed of by spreading on fields for fertilizer in 1962. At that time the value of the plant nutrients was estimated at about 4.0 cents per cwt. Using a ratio of three tons of whey being equal in plant nutrients to a ton of manure, whey was recently worth about 3.3 cents per cwt. for fertilizer. Some research had been done on the application of whey to certain pasture and field crops. The Soil Science Department at the University of Wisconsin is currently conducting research on the amount of whey that can be absorbed in different types of soil and that can be tolerated by different crops.

The same problems in field disposal exist today, as in 1962. Namely, inability of the land to absorb excess quantities of whey, inability of microorganisms and certain crops to effectively convert and use excess quantities of whey, and the coincidental flush in whey production with periods when it is often difficult or impossible to get on the land or for whey to soak into the ground.

The use of whey for fertilizer has been quite limited in Wisconsin. Nutrients contained in a ton of whey had a very recent value of approximately $1. Whey has increased the yields of corn and grass. The large amount of potential nitrogen in whey has not helped legume crops and in some cases has actually decreased yields. Nevertheless, the use of whey for fertilizer has considerable potential, for certain crops, if farmers are apprized of the potential, availability, and cost of the product.

WHEY PROCESSING

No attempt will be made to outline in detail all of the potential methods

by which whey can be utilized (28)(29)(30)(31)(32)(33)(34)(35)(36)(37)
(38)(39). Certainly there is no single solution to the problem; it will require a full gamut of different types of products and processes in order to
divert whey from wastewaters. Some of these products and processes already
exist, whereas others await development. Acid whey, because of the presence of from 0.5 to 0.7% lactic acid, provides problems in respect to drying the material and also in respect to its utilization in food.

Foam spray-drying (30)(40)(41), foam mat-drying (42), reverse osmosis
(43)(44), gel filtration (33), electrodialysis (45), hydrocolloid protein
precipitation (46), ultrafiltration for protein recovery (43)(44), utilization
of the growth of yeast protein (47)(48)(49), fermentation (50)(51)(52)(53),
and animal feed use (54) are all methods which have potential for the conversion of whey into more usable forms.

Membrane Processing

A two-step membrane process has been demonstrated for the treatment of
cottage cheese whey. The process produces valuable protein and lactose
by-products while reducing the BOD of the whey effluent. The process
has been studied in detail in prototype experiments at Abcor, Inc., Cam-
bridge, Mass., and in a 10,000 lbs./day pilot plant at Crowley's Milk
Company, Binghamton, New York, as described by R.L. Goldsmith, D.C.
Goldstein, B.S. Horton and S. Hossain of Abcor, Inc. and R.R. Zall of
Crowley's Milk Company in Proceedings of the Second National Symposium
on Food Processing Wastes, EPA Publication 12060, March, 1971, Washington D.C., U.S. Government Printing Office.

In the two-step process, a protein concentrate is first recovered in an ultrafiltration operation. In the second step, ultrafiltration permeate (depro-
teinized whey) is concentrated by reverse osmosis to provide a lactose
concentrate. The protein concentrate can be further concentrated and/or
dried; and the lactose concentrate can be further concentrated and lactose
recovered by crystallization, or otherwise processed. Alternatively, the
membrane concentrates can be used directly as fluid products.

Operation of the pilot plant was successful and almost troublefree. BOD
reduction of the raw whey was about 97%, from an initial value of about
35,000 milligrams per liter to less than 1,000 milligrams per liter. Membrane life was excellent, and membrane fluxes were economically high.
Membrane flux, membrane rejection, and membrane life for both ultrafiltration and reverse osmosis sections of the pilot plant are discussed in
this paper.

The pilot plant produced protein and lactose products with low total plate
counts, and nil coliform counts. Using the cleaning procedure developed

in the prototype program, total plate counts were typically below 50,000 org./ml., or less than the limit for Grade A milk. Thus, the pilot plant was of a sanitary design and produced dairy grade products.

Capital cost for a 250,000 lbs. cottage cheese whey/day demonstration plant was projected at $610,000 in 1971. This included both the ultrafiltration and reverse osmosis sections of the plant, tanks, and a building to house the plant. Projected operating costs were $196,000 per year. Projected income from utilization of the protein and lactose concentrates makes the process profitable and results in an attractive return on investment.

Figure 58 shows a simplified flow sheet for the two-step whey treatment process. Cottage cheese whey, with or without filtration for fines removal, is introduced into a low pressure UF unit (step 1). In this operation, whey is concentrated ten to thirtyfold by volume. Ultrafiltration membranes are used which retain only the whey proteins. Thus, it is possible to obtain a protein concentrate with a higher proportion of proteins in the dissolved solids, since lactose, nonprotein nitrogen, lactic acid, and minerals pass through the membrane. Operation is typically in the pressure range of 10 to 100 psi, and at temperatures of 60° to 130°F. The protein content of raw whey can be increased from an initial value of about 0.6% up to levels approaching 20% in this step.

Composition of the protein concentrates on a dry solids basis is shown as a function of the degree of water removal in Figure 59. It is apparent that at water removal levels exceeding 95%, protein concentrate streams can be generated with a protein composition of up to 80%. Concentrates of this composition were in fact generated in the course of the experimental program.

The permeate (ultrafiltrate) from the UF unit is introduced into a second membrane step. In an RO operation, this stream is concentrated from approximately 6% solids to 20 to 25% solids. Typical operation would be in the pressure range of 500 to 1,500 psi, and at a temperature of 60° to 100°F. The membrane in the RO section is chosen so as to retain as great a proportion of the organic solutes as possible, resulting in the permeate having a low BOD.

The final effluent from the RO section can either be reused within the dairy or cheese plant or discharged, with or without treatment for residual BOD removal. Where pollution control regulations are not overly stringent, a moderate salt rejection membrane can be used in the RO section to permit partial desalting and lactic acid removal from the lactose concentrate. In general, from the point of view of pollution control, this option will probably not be exercised.

FIGURE 58:   MEMBRANE PROCESS FOR WHEY TREATMENT

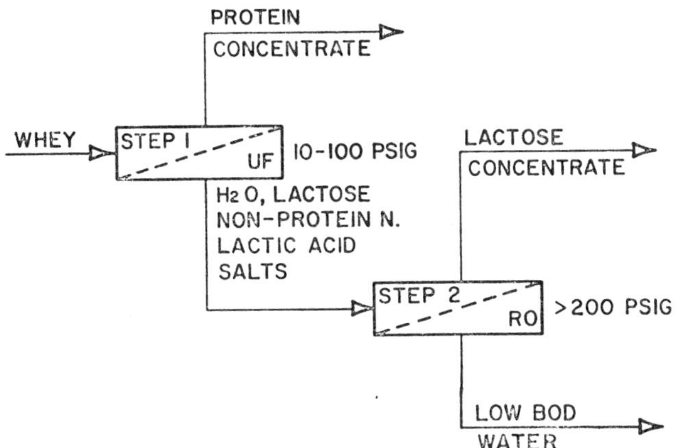

FIGURE 59:   COMPOSITION OF WHEY PROTEIN CONCENTRATE

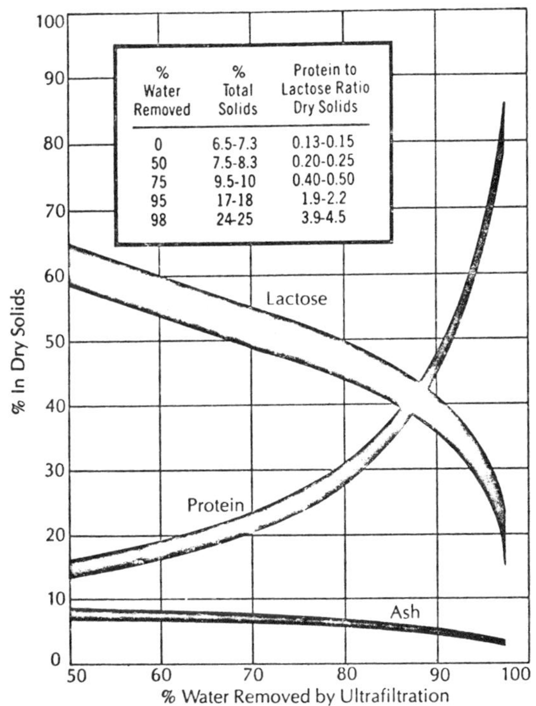

% Water Removed	% Total Solids	Protein to Lactose Ratio Dry Solids
0	6.5-7.3	0.13-0.15
50	7.5-8.3	0.20-0.25
75	9.5-10	0.40-0.50
95	17-18	1.9-2.2
98	24-25	3.9-4.5

Source: EPA Report 12060, March 1971

The protein concentrate can either be used directly by incorporation into food products, or it can be dried. Drying may be preceded by concentration by vacuum evaporation. The lactose concentrate can be further concentrated by evaporation and the lactose can be recovered in a simple crystallization operation.

According to Chemical Engineering magazine for October 29, 1973, page 58, a membrane process developed by Westinghouse can convert whey from a disposal problem to a source of two saleable products.

The process employs untrafiltration and reverse osmosis modules, each consisting of tubular membranes supported by a porous matrix of resin-bonded sand. The ultrafiltration section removed up to 99% of the protein present, which can be processed for marketing as a protein supplement; the reverse osmosis modules then treat the whey to take out as much as 99% of its lactose. With protein and lactose removed, the whey has substantially lower BOD and almost none of the offensive odor usually associated with this stream.

Biological Treatment

Whey recovery by evaporation and drying may well be the most satisfactory solution to the problem of waste whey as pointed out by T.P. Quirk and J. Hellman of Quirk, Lawler & Matusky Engineers of New York City in "Proceedings of the Second National Symposium on Food Processing Wastes", EPA Publication 12060, March, 1971, Washington D.C., U.S. Government Printing Office.

However, as Quirk and Hellman go on to point out, recovery is a fractional proposition that misses about one-fifth of the production which escapes as dilute rinsewater. The treatment of whey-containing effluent liquors by packed tower trickling filter treatment at the Walton, New York plant of the Breakstone Foods Division of Kraftco has been described in some detail by Quirk and Hellman also.

A schematic flow diagram of the recommended wastewater treatment facilities arrived at as a result of this study is shown in Figure 60. The system employs primary settling, two-stage packed tower trickling filters, final settling, coagulation, sterilization by chlorination, and sludge dewatering. Grit removal facilities have not been provided because the raw waste is anticipated to contain virtually no grit, and digesters (requiring grit protection) are not included in the sludge disposal system.

The screened influent will be pumped and metered to the primary settling tank influent distribution chamber and mixed with a recirculation flow from the second stage packed tower. The combined flow will continue through

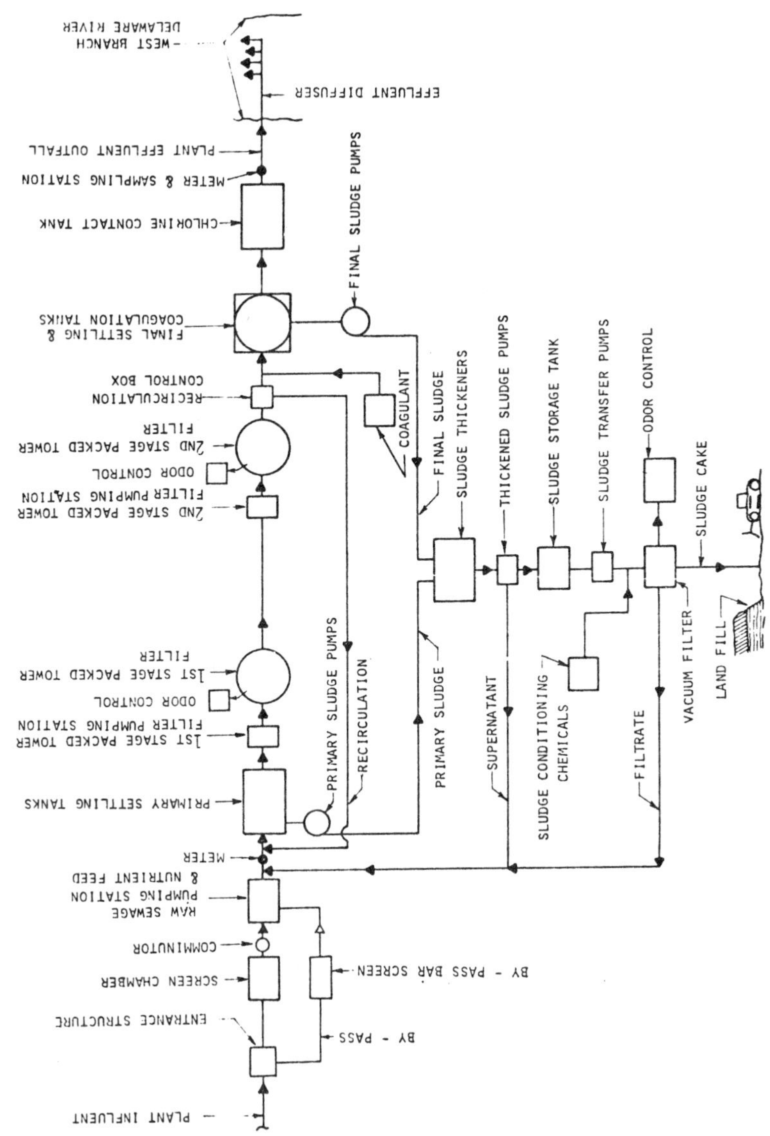

Source: EPA Report 12060, March 1971

the primary tanks. The effluent from the primary settling tanks is pumped up to the first stage packed tower distributor. This first stage effluent is pumped to the second stage packed tower distributor, about 50% of the effluent of the second stage packed tower will be returned as a recirculation flow to the influent distribution chamber of the primary settling tanks. The process flow stream continues through the final settling and coagulation tanks, chlorine contact tank and is finally diffused in the receiving waters.

Sludge from the primary and final settling tanks and coagulation tanks is pumped to sludge thickeners. The sludge thickener underflow will be stored in a sludge storage tank prior to dewatering. The sludge cake will be disposed of in a sanitary land fill.

The treatment of whey bearing waste and its sludge dewatering is associated with objectionable odors. Covers will be provided for the packed towers and odor control ventilation equipment would be provided. An architectural rendering of the complete treatment facilities is presented in Figure 61.

The summary conclusions of Quirk and Hellman's study of the trickling filtration of whey effluent were as follows.

1. Trickling filtration of whey effluent and whey effluent mixed with sewage has been shown to be an effective treatment method both on absolute terms and on a relative basis when compared with activated sludge.

2. A comparison of whey treatability with that of other industrial effluents, demonstrates that whey treatability compares well with that of the average industrial effluent when packed tower trickling filters are employed.

3. Nutrient additions other than ammonia nitrogen were not productive of increased rates of biodegradability. The addition of ferrous iron did not, as anticipated by other investigators, result in an increased treatability.

4. Nitrogen addition resulted in a 40% increase in the rate of BOD removal.

5. The inherently acidic reaction of a whey effluent was shown to have no adverse effect on trickling filtration performance using high porosity media. pH variations from 7.0 to 4.5 were not detrimental but, rather resulted in an increase of BOD removal rate as pH decreased. pH increases above 7.0 were shown to result in a reduction of BOD removal rate. Although these comments are based on limited data, the trend of pH influence appears clear.

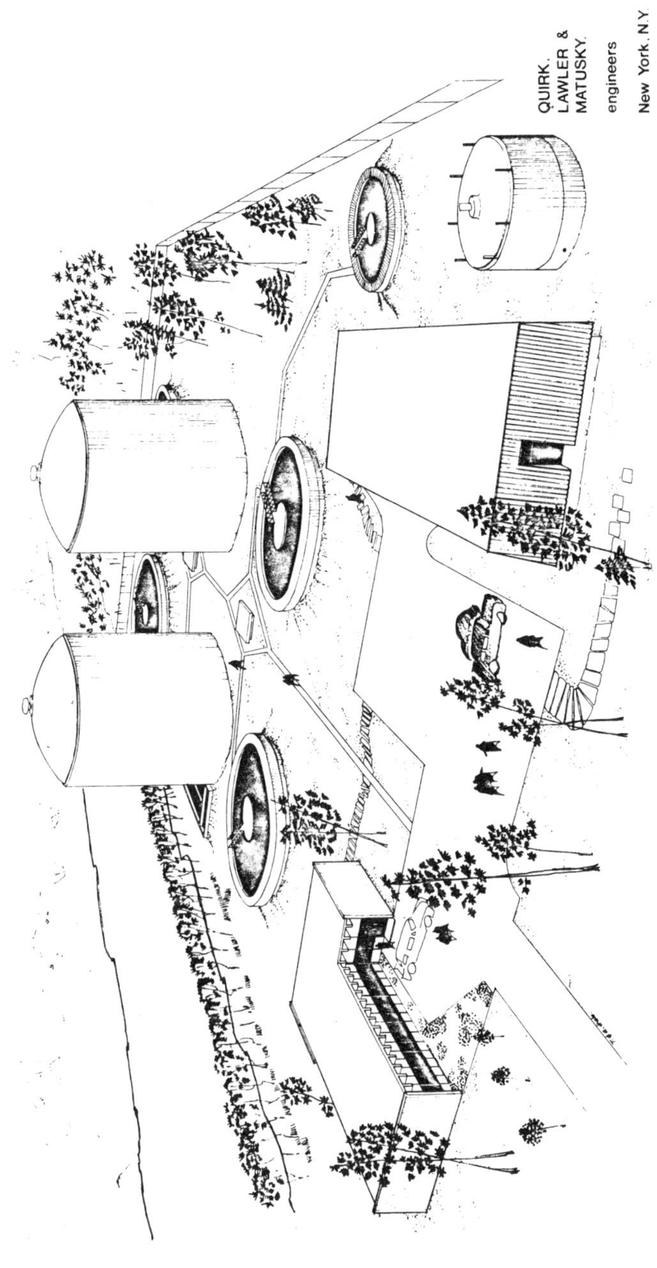

FIGURE 61: WALTON MUNICIPAL WASTEWATER TREATMENT PLANT

Source: EPA Report 12060, March 1971

6. Trickling filter operating variables of temperature, recirculation and hydraulic application rate have been quantified in a verified, process design model. The number of implications which can be drawn from a sensitivity analysis of the model are too extensive to be within the scope of this study.

Several qualitative comments can be made: (a) Increased temperatures are beneficial for process performance. (b) Recirculation should be included only to insure adequate application velocities and not to provide additional removal efficiency. (c) The inclusion of recirculation will require provision for additional filter volumes which will normally outweigh the effects of repeated application. (d) The need to maintain application velocities sufficiently high to insure adequate wetting of packing media will require towers of maximum practicable height or stage treatment to provide high BOD removals, economically. This latter requirement is dictated by media characteristics rather than whey characteristics.

7. Trickling filter performance will be sensitive to flow variation rather than BOD variation. Response times of from 6 to 24 hours should be experienced after a significant change in hydraulic application rate. Under normal operating conditions, filter performance should be stable.

8. Filter sludge growth will be prolific and will require a high porosity media of low susceptibility to retention of sloughed filter slime to avoid plugging.

9. Filter odors should not be offensive at organic loadings requested for high BOD removals. However, filter installation should be provided with covers if proximate to odor sensitive areas.

10. Final sedimentation of filter effluent may require coagulation for production of a low solids effluent and/or for removal of suspended BOD to insure an overall plant performance above 90% BOD removal.

11. Secondary sludge can be thickened using gravity equipment. However, thickener requirements will be significantly greater than that used for domestic sewage.

12. Dewatering of secondary sludge can be accomplished by vacuum filtration. Centrifugation performance would be poor and would not be recommended. Vacuum filtration characteristics of trickling filter sludge exceed, significantly, those of activated sludge. Dewatering requirements may well control the selection of the BOD removal process.

WHEY DRYING

Spray dried whey powder, concentrated whey, and lactose are all products
that may go into human consumption. Roller dried powder and albumin are
usually made into animal feed. Because of the high milk sugar content in
whey, some plants remove a portion of this sugar as crystallized lactose.
The lactose is then purified and used in the drug and candy industries. The
remaining whey is either spray or roller dried.

Since whey also contains a considerable amount of the albumin fraction of
milk, some plants are extracting this albumin and using it in animal feed.
The albumin is extracted from whey by means of a heat and acid process
which causes the albumin to float. It is then skimmed off, packaged, frozen
and used for mink feed. A newer technique removes the albumin directly
from the whey with a centrifuge.

Dried whey is the largest single end use of whey that is processed. The two
methods most common in Wisconsin are roller drying and spray drying. Roller
drying is most common in small plants. The dried whey from roller processing
is not acceptable for human food and, therefore, is usually sold at a lower
price than human grade powder. In some instances, several cheese factories
have combined their whey output and are drying in one location. In other
cases, either because of geographic location or sufficient volume, a plant
may have its own roller plant.

Spray drying is best suited for handling large volumes of whey. Because of
the large investment required, equipment is often utilized that will produce
a product suitable for human consumption. All spray drying plants in Wisconsin have processing capacity that is greater than that of any of the present roller plants. Since there are few cheese factories that are large enough
to efficiently utilize spray drying equipment, this type of installation is
usually in connection with several plants. The processing facility may be
located at one of the factories or at a separate location. If the factories
that form a central drying facility are located considerable distances apart,
it may be feasible to condense the whey before transporting it to a central
drying plant.

In the following discussion of whey drying, considerable price information
is given. These figures are from a University of Wisconsin publication of
July, 1965 (29). No attempt has been made to update these costs, and it
should be borne in mind that costs of all processing components, with the
exception of the whey, have increased considerably since then.

Roller Drying

A number of individual cheese factories, or several cheese factories working together, are now processing their whey into roller dried powder. In this process, the whey is not normally cooled after separation, but is piped directly to a storage tank near the roller dryer. The whey is not concentrated, but is preheated and run directly onto the rolls. This type of whey disposal has an advantage for the small plant. It is quite inexpensive to get started, and if a market can be found, at least a small return may be realized from the whey. Roller dried whey powder is hygroscopic, that is, it will take up moisture from the air and eventually harden.

When the roller dried powder is first taken off the dryer, it is in a lumpy form often described as "popcorn." After this "popcorn" is cooled, it must be run through a hammer mill for pulverization. The ground powder is then bagged and usually shipped to a feed manufacturer for blending into feed. The market price for animal feed dried whey has been quite low for several years because of the large amount of such feed available.

Plants processing whey into human grade powder are often faced with the alternative of holding large amounts of powder for a better market or sacrificing it at animal feed prices. The animal feed market price is usually 3 to 4 cents a pound lower than the human powder market price. For the past several years, prices received by processors of animal feed whey powder have been approximately 5.0 to 6.0 cents per pound as contrasted to 8 to 9 cents per pound for human grade whey powder. The processor of animal feed with a market price of about 5.0 cents per pound must keep processing costs extremely low. However, at this price an efficient roller drying plant may break even or make a small profit.

While roller drying of whey is often a realistic alternative for the small cheese factory, it is not readily adaptable to large operations. The same disadvantages are present as for large roller nonfat dried milk operations.

One problem in large-scale roller drying operations is the number of rollers needed. A plant to process 1 million pounds of raw whey per day would need 13 dryers, each with a capacity of 77,000 pounds of raw whey in 22 hours. The dryers alone would require about 5,000 square feet of floor space. Additional space would be needed for storage, heaters and boilers. The amount of labor needed would be greater than for a spray drying plant of equal capacity.

Three limitations to large size roller operations are that it is: almost impossible to make a nonhygroscopic powder on a roller dryer, much more difficult to make a powder of human grade on roller equipment than with spray equipment and desirable in handling large volumes of whey to partly condense the whey before drying which would add to the original investment. Total investment in a roller drying plant with a capacity of 1 million pounds of raw whey per day would probably exceed $400,000.

Using low cost construction, and as much used equipment as possible, it
was estimated that a whey roller drying plant with a capacity of 77,000
pounds of raw whey in 22 hours could be built for about $46,000. The
equipment cost includes a plate cooler for cooling whey after separation.
This step in the operation is not normally used by plants now processing
whey. However, because of the possible danger of pathogenic or disease
producing organisms being present in whey, it was deemed advisable to
include a plate cooler for cooling whey below the ideal growing temperatures for these organisms. The cooling step is especially advisable if the
whey is to be held some time prior to actual drying.

A building to house the roller dryer was of low-cost cement block construction with a minimum of frills. The warehouse was a wood frame building
with a wood floor. This type of floor is preferable for storage of dried
whey because wood does not "sweat" in warm weather as does concrete.

Fixed costs per day for operating the roller drying plant at capacity were
approximately $67. This includes depreciation, interest, insurance, taxes,
maintenance, labor, and cleanup materials. The largest single item of
fixed cost was labor. Variable cost at capacity was $123 per day. This
cost included an extra labor charge of about $14. Thus, total fixed and
variable cost for operating the plant at capacity was estimated at about
$190 per day. On a unit product basis, the cost would be 3.74 cents per
pound of powder.

Estimated processing costs for this size plant varied from a low of 3.74 cents
per pound of powder in June to a high of 4.89 cents per pound in November,
with a weighted average cost for the entire year of slightly over 4 cents per
pound. If the powder could be sold at an average price of 4.5 cents per
pound, the estimated net returns would be approximately 1/2 cent per pound
of powder, or about 3.2 cents per cwt.

This type of whey drying is quite common in Wisconsin. Several operations
are owned cooperatively by small cheese factories. The whey drying operation may be located at one of the plants, usually the largest, or at a separate location and all of the plants bring their raw whey to the one location
for processing. In areas where alternative methods of disposal have been
costly, roller drying operations have been quite successful.

Spray Drying

The process for spray drying whey differs from that of spray drying skim
milk because of the high lactose content in whey. If whey is spray dried
by the same process as for skim milk, a hygroscopic product results that is
similar to roller dried whey. The powder takes up moisture and becomes
rock-like in consistency. It is then necessary to grind up the powder

before it is suitable for further use.

Two variations from the skim milk drying process have been developed to overcome the instability caused by the high lactose content of whey. The first is keeping the concentrated whey in storage tanks, under continuous agitation, for about 24 hours prior to drying. This procedure allows the lactose to crystallize and will stop the powder from taking up moisture in storage.

The second variation is the use of two-stage dryers. The first stage is similar to conventional vertical dryers for skim milk. However, the powder is taken off the bottom of the first stage dryer at 10 to 12% moisture. The powder is then put through a redryer for final drying to about 97% total solids. The partially dried whey usually moves from the first to the second stage on a continuous belt. This step allows further crystallization of the lactose and the resulting product is a stable, nonhygroscopic powder.

Because of the large investment needed for spray drying equipment, most processors that perform this type of operation are set up to produce a product that is suitable for human consumption. However, the present available capacity for producing whey powder of human grade is more than adequate, and many plants are often forced to sell human grade powder in the animal market.

For this reason, many whey processors have attempted to diversify and to find other whey derived products that have a more reliable market and a higher return than human grade dried whey. One such product is lactose. Most of the large processors of spray dried powder are able to remove at least a portion of the lactose from the whey before final drying. In recent years the lactose market has been stronger than the whey powder market and, consequently, the higher returns from lactose are often able to offset an unprofitable drying operation.

Just as roller drying is limited in handling large volumes, spray drying is an uneconomical way of handling small volumes. Because of the large investment required for spray drying equipment, volumes of 500,000 pounds a day, or higher, are needed if the operation is to be profitable. Costs should be low enough so that if it is necessary to sell in the animal feed market, the dried whey can be disposed of for at least breakeven prices. Large volume operations are needed to reduce costs to this level.

In any operation as large as is necessary in spray drying of whey, diversification is desirable. As noted earlier, in recent years the prices of some whey derived products have been higher than prices of dried whey. Lactose, albumin, yeast and blended products of dried whey and nonfat dried milk are examples. It is also desirable that all products should be suitable for human consumption. Although the human market for whey products has been limited, most operators feel that it is an expanding market and will improve in the future. However, it must be recognized that present plant capacity is more than adequate to supply both the animal feed and the human market for whey powder. Human grade whey products often have trouble finding a market and must be disposed of as animal feed. Nevertheless, any considerable future improvement in net returns to plants depends on being equipped to process human grade products.

Because the market for whey products is often undependable, any large processing plant should have capital adequate to finance a large inventory of finished product. Storage may be necessary for considerable periods of time if market conditions are not right. Since whey production parallels cheese production, the largest demands on processing and warehouse space will be in the spring and early summer. The ability to store finished products until supplies are short, normally in the fall, may mean the difference between disposing of inventory at lower rather than higher prices.

Many of the operations connected with whey processing are covered by patents. For this reason, any organization that is considering any whey processing should become familiar with which processes are patented. Arrangements may have to be made with the patent holders for lease or royalty rights. It is much safer to have patent clearance before starting an operation than be faced with a patent infringement suit at a later date. The reader is referred to <u>Cheesemaking Technology</u> by M.E. Schwartz, Park Ridge, New Jersey, Noyes Data Corp. (1973) for patent information on whey processing.

Estimated costs for spray drying whey were determined by synthesizing two model plants. Plant 1 has a theoretical capacity of 1 million pounds of raw whey in 21 hours. Plant 2 had an assumed capacity of 500,000 pounds of raw whey in 17 hours. The remaining time was to be used for cleanup. The buildings and equipment for each of these plants is assumed to be of such quality that the products manufactured could be used for human consumption.

The cost of Plant 1 was estimated at slightly over $1 million and Plant 2, about $700,000. While the cost of Plant 1 was 40% more than Plant 2, its capacity was almost double that of Plant 2. As is true with most large processing equipment, once the initial investment is made, a much smaller proportionate investment is required to increase the capacity.

The estimated costs for these plants were based on actual costs of operating whey processors, estimated costs of equipment by manufacturers and estimates of the Department of Dairy and Food Industries, University of Wisconsin. This type of cost estimate involves accounting information and engineering data. These are combined so as to duplicate actual operating conditions and eliminate variables, such as wage rates, different depreciation schedules and seasonal swings in production. Costs are budgeted for different seasonal levels of production and then combined to give an estimated annual average cost of production.

The fixed costs for Plant 1 were estimated at about $700 per day, and variable costs at about $800 per day. Total daily fixed and variable costs for processing 1 million pounds of raw whey were estimated at about $1,500. Fixed costs for Plant 2 were estimated at about $500, variable costs at about $400, and total fixed and variable costs about $900 per day.

At theoretical capacity, the estimated costs per pound of solids processed is 2.35 cents per pound of solids for Plant 1, and 2.86 cents for Plant 2. Whey powder was assumed to be 97% total solids. On a powder basis, these costs would be 2.28 and 2.77 cents, respectively.

At levels of production below 700,000 pounds, Plant 2 (theoretical capacity of 500,000 pounds per day) can produce whey powder at a lower cost than Plant 1. At volumes above 700,000 pounds, Plant 1 is the more efficient. The estimated costs for Plants 1 and 2 are summarized as follows.

	Plant 1	Plant 2
Estimated total construction costs	$1,000,000	$700,000
Theoretical capacity per day, lbs.	1,000,000	500,000
Estimated fixed costs per day	700	500
Estimated variable costs per day*	800	400
Estimated total costs per day*	1,500	900
Estimated costs per pound of solids	.0235	.0286

*At theoretical capacity

Because of the seasonality of production, plants are not able to operate the year around at capacity. Following the seasonal swing in production, the weighted average annual costs for a plant with a volume of 1 million pounds per day in the flush were estimated at 2.69 cents per pound of solids or about 2.86 cents per pound of powder. This is the processing cost only. Any transportation costs will have to be added to this.

The processing costs cited for Plants 1 and 2 assume that all the whey for processing is already at the plant. Since there are few, if any, cheese factories in Wisconsin that would have a million pounds of whey available for processing, the cost of moving whey from outlying cheese factories to a central processing plant must also be considered.

In this analysis, the assumption was made that whey would be received 12

hours a day from 10 different cheese factories located within 50 miles of
the central plant. These factories were assumed to have a volume of
100,000 pounds per day each, during the flush. The whey could be transported by tractor trailer combinations carrying a load of approximately
50,000 pounds of raw whey on each trip. These tractor trailer combinations
were assumed to cost about $28,000 each. The variable cost was estimated
at 10 cents per mile.

On this basis, the costs for hauling whey from the 10 cheese factories to
the central drying plant during June were estimated at about 1.2 cents per
pound of solids, and during November 1.5 cents with an average cost for
the two months of 1.4 cents.

As was stated previously, no attempt has been made to update the processing
costs that were estimated in 1965 (29). Costs of all processing components,
with the exception of the whey, have increased considerably. For example,
since 1962, construction costs are up 53%, machinery and equipment costs
up 25%, average hourly earnings up 44%, and contract construction up
73%. All of this has occurred while the price for dried whey has remained
relatively constant.

As noted, roller drying is not as common as in 1961, and more of the dried
whey product qualifies as being eligible for human consumption. Spray
drying continues to be the most important drying method, but new innova-
tions in drying and processing are gaining in importance. Dried whey is
still a product of last resort and one of small profit. Whey proteins, lactose,
blended and specialty products are more profitable than dried whey.

WHEY MARKETING

Large whey processors usually sell their products directly to the end user
through their own sales organizations. Smaller processors and those affili-
ated with regional organizations usually sell through the latter's sales
organizations, and in some cases through brokers.

The overall market for whey products has been expanding. New uses have
been developed, both as a food and for certain industrial processes. There
is more interest in breaking whey down into component parts and then using
the parts in other foods.

While a whey market as such does not exist, some estimates of relative
return can be made. Processors indicate that the blended products and
those using component parts of whey are more profitable than just dried
whey. However, the returns from dried whey are often better than other
disposal methods. For example, whey drying may be a breakeven operation,

while the next alternative may result in a net loss. At the present time there
there is a lack of uniform standards for use throughout the whey industry.
This may work to the advantage of some processors but to the overall dis-
advantage of the industry, especially in estimating the volume of different
uses of whey, or when potential buyers wish to compare whey products.
Hopefully this situation will be corrected in the next few years.

Whey Prices

Prices of dried whey (animal feed) have been relatively stable for the past
15 years. In 1955, the price of animal feed was 5.53¢/lb. in Chicago; in
1970 the price was 5.68, and in 1971, 4.99¢/lb. (Table 59). During the
same period the price of nonfat dried milk was stable until 1966; from 1966
to 1971 the price more than doubled. Prices for human grade whey powder
have not been generally reported until the last few years. Prices in eastern
areas for human grade dried whey were generally 2 to 3¢/lb. above the price
of animal feed from 1965 to 1970 (Table 59). Monthly average prices in
Wisconsin are shown in Table 60. The difference in price between animal
feed and edible whey is shown in Table 61. Spray animal feed ranged from
less than 1¢/lb. difference to slightly over 2¢/lb. The differences were
greater between roller dried and edible whey. The edible whey ranged from
about 115% to over 200% of the animal feed prices.

TABLE 59:   NONFAT DRIED MILK AND WHEY PRICES

	NFD Milk ¢/lb., fob Factory	Dried Whey ¢/lb., Wholesale Trucklots	
		Animal Feed Chicago	Human Food Eastern Areas
1955	15.35	5.53	
1956	15.24	5.68	
1957	15.29	5.51	
1958	14.09	5.61	
1959	13.60	5.60	
1960	13.66	5.60	
1961	15.45	5.26	
1962	14.79	5.25	
1963	14.43	5.42	
1964	14.62	5.35	
1965	14.66	5.18	7.89
1966	18.19	6.15	9.12
1967	19.92	6.02	9.25
1968	22.36	5.02	8.68
1969	23.50	5.28	8.05
1970	27.3	5.68	7.83
1971	30.7	4.99	

Source: various issues of The Dairy Situation

TABLE 60:   MONTHLY AVERAGE PRICES FOR DRIED WHEY AND NON-FAT DRIED MILK, SELECTED MONTHS, WISCONSIN, 1970 TO 1972

Date	Dried Whey (Human) Spray	Dried Whey (Animal Feed) Spray	Dried Whey (Animal Feed) Roller	N-F Dry Milk High Heat	N-F Dry Milk Low Heat	N-F Dry Milk Grade A
1970						
June	6.67	5.26	4.38	27.74	28.25	-
July	6.63	5.52	4.41	27.74	28.25	-
Aug.	6.52	5.50	4.48	27.68	27.77	28.91
Sept.	6.40	5.56	4.68	27.68	27.68	28.93
Oct.	6.63	5.73	4.86	27.68	27.68	28.93
Nov.	6.81	5.78	4.90	27.68	27.68	28.93
Dec.	6.86	5.88	4.90	27.68	27.68	28.93
1971						
Jan.	6.53	5.25	4.79	27.68	27.68	28.93
Feb.	6.31	4.84	4.45	27.58	27.64	28.93
Mar.	6.08	4.75	4.14	27.55	27.66	28.93
Apr.	6.03	4.61	3.94	31.89	31.89	33.43
May	6.00	4.53	3.75	32.05	32.05	33.43
June	6.00	4.53	3.66	32.05	32.05	33.43
July	6.70	4.63	3.63	32.05	32.16	33.43
Aug.	6.88	4.81	3.64	31.71	32.20	33.43
Sept.	6.88	5.00	3.36	31.50	32.08	33.41
Oct.	6.88	5.00	3.78	31.50	32.08	33.43
Nov.	7.13	5.00	4.07	31.50	32.08	33.43
Dec.	7.18	5.33	4.39	31.63	32.08	33.43
1972						
Jan.	7.25	5.38	4.48	31.72	32.16	33.43
Feb.	7.25	5.63	4.50	31.75	32.25	33.43
Mar.	7.33	5.68	4.51	31.68	32.24	33.38
Apr.	7.23	5.50	4.58	31.63	32.23	33.35

Source: Monthly Cold Storage Report, Wisconsin Department of Agriculture

TABLE 61:   DIFFERENCE IN PRICES BETWEEN ANIMAL AND EDIBLE GRADE WHEY POWDER, WISCONSIN, 1970 TO 1972

	Cents per lb. that Edible Whey Exceeds Animal Feed:		Price of Edible Whey as a Percent of Animal Feed:	
	Spray	Roller	Spray	Roller
1970				
June	1.41¢	2.29¢	126.81%	152.28%
July	1.11	2.22	120.11	150.34
Aug.	1.02	2.04	118.54	145.54
Sept.	.84	1.72	115.11	136.75
Oct.	.90	1.77	115.71	136.42
Nov.	1.03	1.91	117.82	138.98
Dec.	.98	1.96	116.67	140.00
1971				
Jan.	1.28	1.74	124.38	136.32
Feb.	1.47	1.86	130.37	141.80
Mar.	1.33	1.94	128.00	146.85
Apr.	1.42	2.09	130.80	153.04
May	1.47	2.25	132.45	160.00
June	1.47	2.34	132.45	163.93
July	2.07	3.07	144.71	184.57
Aug.	2.07	3.24	143.04	189.01
Sept.	1.88	3.52	137.60	204.76
Oct.	1.88	3.10	182.01	182.01
Nov.	2.13	3.06	142.60	175.18
Dec.	1.85	2.79	134.71	163.55
1972				
Jan.	1.87	2.77	134.76	161.83
Feb.	1.62	2.75	128.77	161.11
Mar.	1.65	2.82	129.05	162.53
Apr.	1.73	2.65	131.45	157.86

Source: Calculated from data appearing in the Monthly Cold Storage Report of the Dairy and Poultry Market News issued by Federal-State Market News Service, U.S.D.A.

FUTURE TRENDS IN WHEY HANDLING

The amount of whey available for processing is bound to increase as the proportion of, and the absolute volume of milk used in cheese, increases. In 1955, about 25% of the total U.S. milk supply was used in butter and 11% in cheese. By 1965, butter production utilized 23% and cheese about 13%, and in 1969, butter utilized 21% and cheese over 15%. In 1971, butter used only about 3 billion pounds more milk than cheese: 21 billion pounds into cheese and about 24 billion pounds into butter. If this trend continues, and at this time there is little reason to doubt that it will, the percent of milk used in cheese will exceed that used in butter within the next two to three years.

The U.S. Department of Agriculture has estimated that there will be an additional 750 million pounds of whey solids within the next few years. This will mean a doubling of the size of the present whey processing industry.

Production of butter in Wisconsin has been steadily declining as plants have shifted out of butter powder and into cheese. In 1969, the production of dried whey in Wisconsin exceeded the production of nonfat dried milk for the first time (Table 62). Milk production has been declining in some areas and increasing in others. Consequently, production is being concentrated in certain parts of the country, primarily the major dairy areas in the East, Midwest, and West. For example, Wisconsin produced over 17% of the nation's milk supply in 1970, and will probably produce 19 to 20% by 1980.

TABLE 62: PRODUCTION OF NONFAT DRIED MILK AND DRY WHEY, WISCONSIN, 1960 TO 1970

Date	Dry Skim Milk for Human Use	Dry Whey
	('000 Pounds)	
1960	424,938	87,495
1961	436,733	72,046
1962	510,206	110,588
1963	469,607	136,433
1964	468,158	136,433
1965	415,611	158,802
1966	272,470	207,789
1967	287,578	222,153
1968	253,538	240,577
1969	208,830	244,202
1970	186,677	292,981

Source: various issues of Wisconsin Agricultural Statistics, Wisconsin Statistical Reporting Service

By 1980, pressures to use whey in an economical manner and not treat it as waste or sewage will increase. This will take the form of more rigid regulations for land and sewage disposal, and more research for new whey products, both industrial and food. There will likely be more regulation and standardization of whey products than there has been in the past. Better reporting of utilization will be required, both in complete products and in products that use components of whey. This should help processors in adjusting to changes in consumption trends and in other aspects of the market.

Much of the economic incentive to use whey in food products will depend on the nonfat dried milk market. The higher the price of nonfat in relation to dried whey, the more pressure to find ways to substitute whey products for nonfat. At the present time, nonfat is 3 to 4 times the price of whey, and in many bakery processes whey and whey blended products are a very satisfactory nonfat substitute.

Forces outside the market that can influence whey utilization include the import and export policies of the U.S. and other dairy nations, such as changes in quotas for imported cheese into this country, government subsidies for exporting dairy products, and the policies of foreign governments. Increasing world demand for quality protein can lead to the development of new food products. Hopefully they can be exported to food deficient areas and still be produced for a profit in this country.

The market for component parts and blends of whey and other products will continue to increase. While dried whey will still use a large part of total whey production, it will continue to be less profitable than selling other products. Competition for the dried whey market will continue to be tough. The firms that can develop and market unique and innovative products have the potential to make the most money.

Some industry leaders feel that a whey bias still exists. The image of whey is such that many people feel it is a disposal problem or "hog feed" rather than a food. Other people feel that the only bias against whey has been in the price, and that if the price is favorable, compared to alternative products, processors will not hesitate to use whey.

Greater cheese and whey production will mean that a market will have to be found for more whey cream. In 1970, the production of American cheese in Wisconsin was about 650 million pounds. Assuming a yield of 0.3 lb. of fat removed from 100 lbs. of whey, the butterfat from whey cream can be estimated at about 17 million lbs. or 21 million lbs., butter equivalent. This is equal to more than 10% of the total butter produced in Wisconsin. Since whey cream also comes from other cheese whey, the above estimate is low.

In summary, the marketing of whey has changed considerably in the last ten years. Most notable has been the greater use of whey in some form of processed product and the decline in the amount of whey that is just dumped or wasted.

Prices for dried whey have been remarkably stable for many years. Those products that use whey components as a part of other foods have been, and will continue to be, more profitable than dried whey. Competition for markets will be as keen as in the past as more whey is processed and seeks a profitable market.

Whey production will increase as cheese production increases. This will keep pressure on the industry to develop new uses, and to use existing methods of disposal more efficiently.

DAIRY PLANT WASTE TREATMENT

Most dairy plant wastes respond to the biological treatment approach. The wastes are similar to municipal wastewater but considerably more concentrated and more readily degraded. The biological treatment approach most generally represents the best and most economical method of treating dairy wastes to reduce the BOD and COD concentrations to acceptable ones. Biological treatment of dairy wastes is, of course, widely provided in municipal treatment systems. Dairy plants also use the biological approach to treat wastes for stream discharge.

Irrigation of the process wastes as a means of disposal is a viable approach for some dairy plants. The common methods of irrigation are: spray fields, spreading and ridge and furrow application. The plant has to own or have under lease adequate land of suitable type to take the volume of wastewater to be disposed of without runoff into the streams of the area. Further, the irrigation land must be close enough so it can be reached by pipeline or truck on an economical basis.

Some isolation is necessary for a satisfactory irrigation site so no odor nuisance conditions will be created. Further, care must be excercised in application so ponding on the land does not take place. Irrigation works best in areas where the winter climate is not severe but can be used in such wintry areas if properly operated. Some lagoon holding space, in some cases with the use of aeration, is desirable in conjunction with the operation of irrigation projects to provide holding flexibility for rainy or wintry periods.

Spreading on land represents another method of discarding whey for certain types of plants. There are significant numbers of small cheese plants which are somewhat isolated and cannot afford concentration equipment and stay in operation. Further, as a result of the volume of whey produced and the scatter of the plants in the particular region central drying facilities cannot be justified.

Under these conditions, in order for these plants with their vital cheese production to remain in operation, land disposal of the whey must be practiced. In many areas a mutually beneficial arrangement has been worked out with the farmers in the immediate vicinity of the plant to dispose of the whey on their farms to take advantage of the nutritional value to benefit the land.

Dairy food plant wastes are treated primarily by biological oxidation methods. Biological oxidation is a function of the microflora of the waste treatment system, which in turn is dependent upon the composition of the wastes. According to Dias and Bhat (55), the dominant microflora in a municipal activated sludge plant in decreasing order are Zoogloea, Comannanmonas, Pseudomonas, Micrococcus, Flavobacterium, Achromobacter, Alcaligenes, Corynebacterium, Bacillus, Spirillum and yeasts. All of these have been found in activated sludge of dairy waste treatment.

However, the dominant microflora reported by various investigators for dairy activated sludge are quite different from that of domestic activated sludge. Adamse (56) found that the adaptation of the bacterial flora from municipal to dairy wastes required approximately 50 to 60 days. In his investigations, the dominant microflora were Corynebacterium, of the Arthrobacter type, followed by Achromobacter. Pipes (57) in his investigations of activated sludge plants found that the dominant microflora in his particular investigations were Bacillus, Beggiatoa, Arthrobacter, Actinomycetales, and Sphaerotilus. Similar microorganisms have been reported in the microflora from trickling filter and aerated lagoon systems operating on dairy wastes.

Only limited investigations have been made of the metabolisms of the microflora of various dairy food plant waste treatment systems. Since the dominant microflora of dairy food plant waste treatment systems appear to differ from the microflora of municipal systems, there is need for further specific information. The most exhaustive study in respect to the factors affecting the metabolism of the microflora of dairy food plant waste treatment systems has been conducted by Adamse (56). In investigation of the growth characteristics of the major groups of organisms in his system, he found that the Arthrobacter species was the most active and efficient of the three types, with a generation time of 2.5 hours at 15°C.

In batch experiments with well aerated activated sludge, the oxygen demand immediately after feeding artificial dairy wastes was found to exceed the oxygen supply. This caused a decrease in dissolved oxygen content of the

material and a simultaneous increase in the COD of the effluent. During
the deterioration of the substrate, acid intermediates from carbohydrate
breakdown were excreted to the system, with sharp drop in pH. This suggested that the oxidation of the carbohydrate fraction was limited even
though adequate amounts of oxygen were supplied. The decrease in the
pH proceeded more slowly when the dissolved oxygen content approached
0, indicating that acid formation was due to activity of aerobic organisms.

From the results obtained, the conclusion was reached that the dispersion
period for protein fraction proceeded at a remarkably lower rate than dispersion of carbohydrate fraction. Typical data are presented in Figure
62. In studying the metabolism of sludge microorganisms, Adamse found
that the drop in pH in the activated sludge shortly after feeding was due
to the accumulation of acetic acid in the presence of adequate dissolved
oxygen, and lactic acid accumulated when dissolved oxygen was below
0.5 ppm. Both acids were degraded after lactose had been exhausted, but
lactic acid was more readily oxidized than acetic acid.

FIGURE 62:   DIAGRAMATRIC REPRESENTATION OF CHANGES IN
             DAIRY WASTEWATER DURING TREATMENT OF ACTIVATED SLUDGE

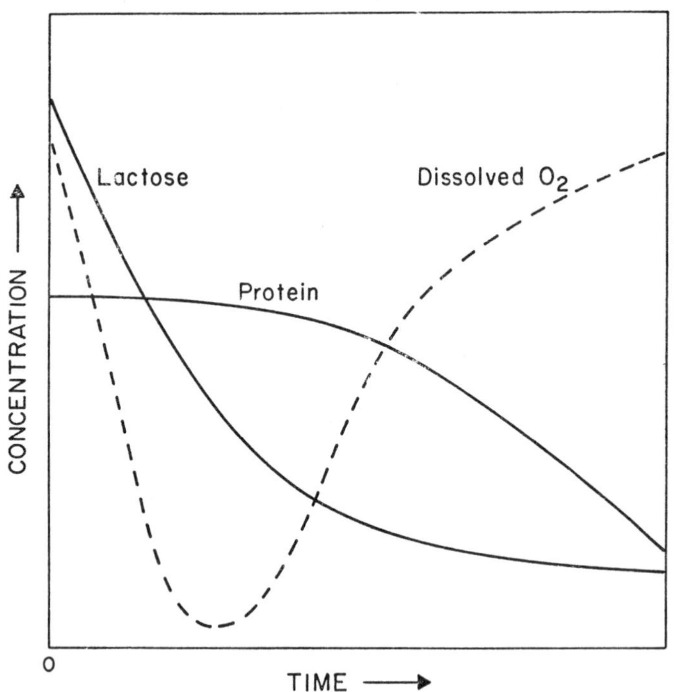

Source: Harper and Blaisdell, Symposium on Food Processing Wastes

The apparent degradation rate of dairy wastes was found optimal at pH 6.5
of the activated sludge suspension. Of the lactose added to the activated
sludge, approximately 10% was respired, up to 50% was accumulated in
polysaccharides in the cell biomass and the remainder was used for the syn-
thesis of cell constituent and partly accumulated as intermediates in the
sludge mass.

The difficulty in the treatment of whey appears to be related to the inability of microorganisms to degrade whey protein and the ability of these
proteins to interact with the polysaccharide of the cell mass. Incorporation of whey or whey wash water in the activated sludge treatment plants
has been known to cause excessive foaming. The foam has been shown
by earlier studies to be a combination of microbial cells and major whey
protein B-lactoglobulin. Although the BOD/N ratio of milk wastes approximates those desired (20:1), the available nitrogen is considerably less
because of the binding of the protein to the cell mass and the inability of
enzymes to readily degrade material.

Ammonium nitrogen has been frequently indicated as a requirement in dairy
food plant wastewaters. However, many dairy food plant waste treatment
systems and some investigators have failed to show an increased efficiency
in BOD reduction through the addition of nitrogen; and the exact reason
for these observations needs further investigation.

The types of biological waste treatment methods utilized in the dairy food
industry are primarily activated sludge, trickling filters, aerated lagoons,
irrigation and a combination of these. Characteristics with respect to
loading and efficiency of BOD reduction are summarized for selected treatment systems in Table 63. The data are not inclusive of those collected,
but illustrate the differences that have been reported.

Efficiencies of biological oxidation treatment systems for dairy wastes are
generally reported to be higher than 90%. In the site visits and from those
studies that have been exhaustively reported in the literature, it becomes
apparent that this degree of efficiency is achieved only for a part of the
operating time. Based on some 20 evaluations of treatment plants for dairy
wastes, it was found that the waste treatment facilities were less than 70%
efficient about 25% of the time. Figure 63 shows typical frequency plots
of the variability of waste treatment for several systems treating only dairy
wastes.

Because of the high BOD load, in many modern plants up to 4,000 ppm
even 99% efficiency does not bring the BOD in the effluent down to a level
acceptable in some states for direct discharge to a stream. As a result, an
increasing number of plants that have their own treatment facilities are combining two or more biological treatment methods. It would appear that a

TABLE 63:   COMPARISON OF WASTE TREATMENT METHODS FOR DAIRY FOOD PLANT

Method of Treatment	BOD$_5$ in ppm				% Reduction In BOD$_5$		Loading		Retention Time (hrs)	
	Untreated		Treated							
	Range	Aver.	Range	Aver.	Range	Aver.	Range	Aver.	Range	Aver.
Activated Sludge (10)	84–2,920	404	7–632	85	17–99	79.08	0.018–1.10	0.35[a]	----	----
Activated Sludge	180–15,000	1903	1–1700	163	57–99.8	88.7	0.06–-485	137[b]	1.5–120	21.4
Trickling Filter	130–32,000	2846	10–1400	81'	35–99.5	77.4	0.4–-49.4	12.7[c]	No information avail, estimate 36–48 hrs.	
Aerate Lagoons	75–10,000	286	25–800	191	52–89.5	76	10–395	162[d]	168–1080	420 (17 days)
Irrigation[e]	–	220	–		–		2500–90,000	27,700[f]	–	–

[a]lb BOD/lb MLSS; [b]lb BOD/100 cubic feet; [c]lb BOD/cubic yard; [d]lb BOD/day/acre; [e]BOD loading generally not given, the value of 220 is given as maximum loading rate recommended; [f]volumetric loading in gallons/day.

Source: Harper and Blaisdell, Symposium on Food Processing Wastes, March, 1971

FIGURE 63:   VARIATION IN THE EFFICIENCY OF DIFFERENT TYPES OF WASTE TREATMENT FOR DAIRY WASTE (AL = AERATED LAGOON; AS = ACTIVATED SLUDGE; TF = TRICKLING FILTER)

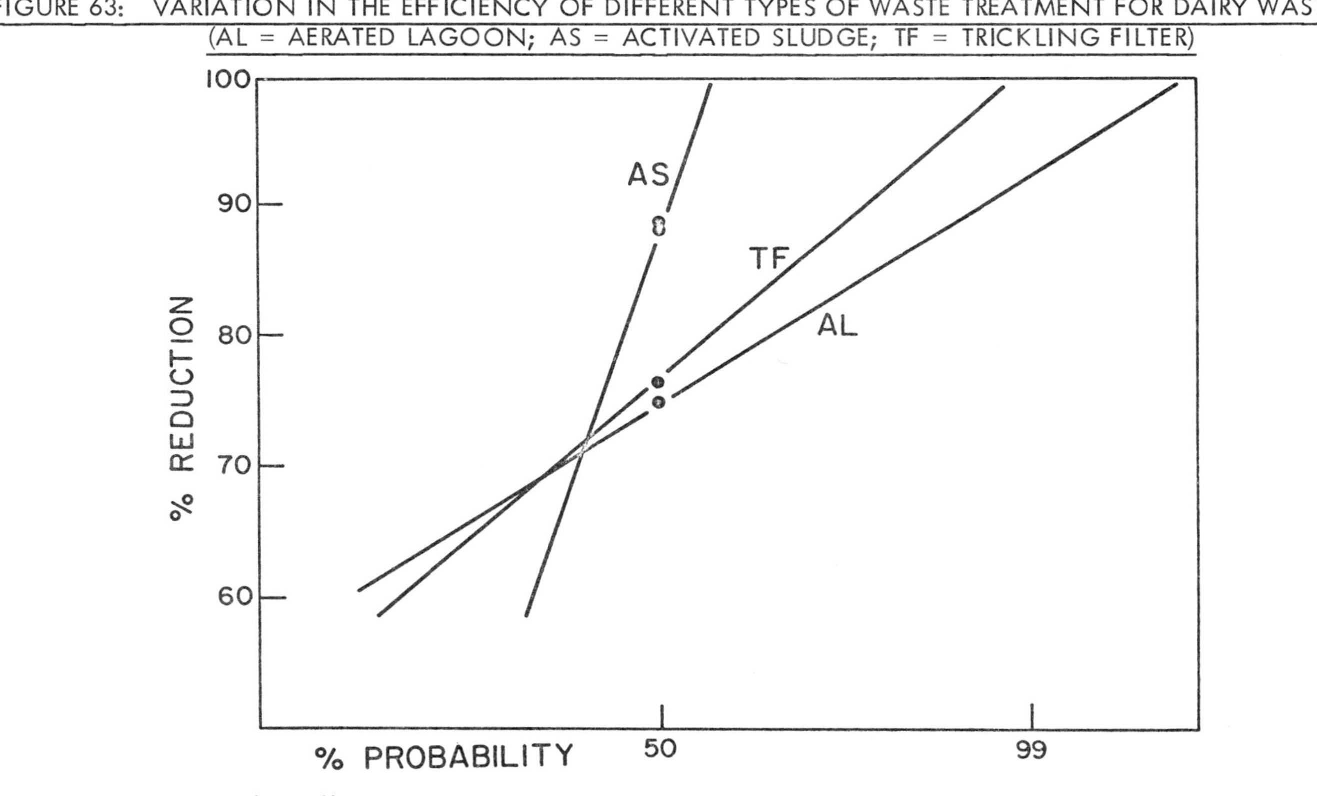

Source: Harper and Blaisdell, Symposium on Food Processing Wastes, March, 1971

two-stage or three-stage biological oxidation is necessary in order to provide for optimum treatment to meet stream standards.

The relative efficiency for BOD and COD treatment under commercial dairy
plant operations obtained from plants surveyed in this study are presented
in Table 64. Values show that in one dairy food plant waste in which the
BOD of the effluent was 15 ppm, the COD of the effluent was 175 ppm.
This may reflect the refractory nature of certain detergents and chemicals
used in the cleaning and sanitizing of dairy food plant equipment. Problems encountered in the treatment of dairy food plant wastes (15) include:

1. Periodic poor performance, and lack of correlation
   between loading rate, sludge characteristics and BOD
   reduction.
2. Poor settling characteristics of sludge in activated sludge
   systems, with problems of pin point floc, defloccula-
   tion and bulking.
3. Periodic loss of the biomass in trickling filter systems.
4. Low levels of available nutrients, caused by possible
   complexing of milk constituents with microbial cells,
   and possible feed back inhibition of B-galactosidase
   by lactic acid.
5. Slow recovery of the biological activity of the treatment
   systems, after shock loading, or periods of low feeding.
6. High organic concentrations, high hydraulic and nutrient
   variability and consequent shifts in the microflora of the
   waste treatment facility.
7. Possible "toxicity" in some dairy wastes, attributable to
   sanitizers and detergents.
8. Stable, high strength foams in activated sludge systems
   in the presence of whey or often cottage cheese wash
   water in addition to light froth foams associated with
   over-aeration.
9. Relatively high COD levels in treated wastewaters, with
   BOD-COD ratios of less than 0.1 in some cases.
10. Frequent clogging of diffuser-type aerators and sludge
    breakup.
11. Poor sludge filtration characteristics.

PROCESS SELECTION OBJECTIVES

In the selection of a wastewater treatment alternative, a number of factors
must be evaluated prior to making a final choice as described by Boyle &
Pockowski (58). Among those objectives which most often dictate process

TABLE 64: BOD5/COD RELATIONSHIPS IN RAW AND TREATED DAIRY PLANT WASTES FROM SURVEY OF INDUSTRIAL DAIRY PLANTS

Products Manufactured	Raw Dairy Plant Wastes			After Primary Treatment			After Secondary Treatment		
	BOD_5	COD	BOD/COD	BOD_5	COD	BOD/COD	BOD_5	COD	BOD/COD
Milk Cottage Cheese	3300	3379	.97	---	---	---	---	---	---
Milk Cottage Cheese	3700	7071	.52	145	400	.36	15	100	.15
Milk Yogurt Cottage Cheese	750	850	.88	400	600	.67	80	250	.32
Cottage Cheese Milk	2863	6500	.44	---	---	---	266	700	.38
Milk Cottage Cheese Ice Cream	3200	9000	.41	145	430	.33	15	175	.08
Milk Ice Cream Butter Cottage Cheese	2168	5300	.40	140	350	.40	35	250	.14
Milk Ice Cream Cottage Cheese	3750	9000	.41	---	---	---	150	560	.26
Milk Ice Cream Cottage Cheese	3200	4150	.51	---	---	---	---	---	---
Ice Cream	140	350	.40	---	---	---	35	250	.14

Source: Harper and Blaisdell, Symposium on Food Processing Wastes, March, 1971

selection are (a) effluent criteria, (b) site limitations, (c) wastewater characteristics, (d) wastewater variation, (e) expected life, and (f) cost to
treat. A brief outline of these objectives will precede a more extensive
discussion of specific treatment alternatives.

Effluent Criteria

The requirements for effluent quality from the dairy industry have been
most recently released by U.S. EPA in August, 1972, based upon two comprehensive studies of the dairy industry: "Study of Wastes and Effluent Requirements of the Dairy Industry" by A.T. Kearney & Co. (59), and "Dairy
Food Plant Wastes Treatment Practices" by Ohio State University (2). Currently, these requirements based on "best practicable control technology
currently available" cover only the wastewater parameters of BOD and suspended solids. For new plants under construction or existing plants now
beginning abatement programs, effluent BOD and suspended solids concentrations of 30 mg./liter are expected regardless of influent characteristics
unless there are unusual or restrictive characteristics.

Public Law 92-500, the Federal Water Pollution Control Act Amendments
of 1972, further states that even a higher level of treatment will be required
by July 1, 1977, in those areas where secondary treatment will not meet
water quality standards. Water quality standards must be achieved in all
instances and every evidence indicates that these standards will be whole
body contact recreation plus fish and aquatic life standards. It is likely,
therefore, that requirements for the dairy industry will become more stringent in years to come. Furthermore, additional water quality parameters
such as phosphorus and nitrogen will undoubtedly be added to the list.

Restrictions on phosphorus removal have already become a reality in many
parts of the country. In Wisconsin industrial discharges in excess of 8,750
pounds of total phosphorus per year must achieve 85% removal if discharged
to the Lake Michigan watershed. It is apparent, then, that the dairy industry will have to be concerned about its phosphorus discharges in the near
future. In process selection, therefore, it is important to consider flexibility
and versatility of the flowsheet in providing reliable and consistent effluent
quality. One must consider design based upon both expansion of production and increased restriction of pollutant discharge.

Site Limitation

In the selection of either pretreatment or complete treatment facilities, the
constraints of the site often play a controlling role. Low first cost processes
normally require substantial land areas and may restrict development within
1/4 to 1/2 mile of the facility. Urban locations may be restricted by odor,

noise, or esthetic regulation. Cost to pump or otherwise transport waste-waters to remote locations should not be overlooked.

Wastewater Characteristics

Wastewaters from dairy plant processes normally include substantial concentrations of fats, milk proteins, lactose, some lactic acid, minerals, detergents, and sanitizers. Normally, a major fraction of the pollutants is in a dissolved organic and inorganic form not susceptible to plain sedimentation or flotation. The strength and quality of wastewater varies widely even within plants producing the same dairy food products. That plant management plays an important role in these characteristics is brought out by the Ohio State Report (2). A complete compilation of wastewater characteristics is presented in this document.

The wastewaters from most dairy food operations are substantially higher than domestic wastewaters as measured by BOD, COD, total organic carbon and volatile solids. The carbon to nitrogen ratio of dairy food processing wastewaters is normally higher than that of domestic waste as is the carbon to phosphorus ratio. Thus, processes ordinarily acceptable for handling domestic wastewaters may require considerable modification for handling dairy process wastewaters. Furthermore, combined treatment of dairy plant wastes with municipal wastes may be substantially influenced by the proportion of dairy plant to municipal waste flows.

Wastewater Variation

Discharges of dairy wastewater are often batch or slug dumps producing wide variation in flow and quality. Treatment processes may be sensitive to either qualitative or quantitative shock loads caused by these variations. It is important to evaluate treatment performance reliability in light of these expected variations. Equalization, neutralization, or other forms of load equalization must be considered in concert with the treatment processes which are sensitive to widespread variation.

Expected Life

The treatment facility being designed by industry normally has a short design period owing to the uncertainty in production forecasts. Currently, this policy would seem to be of considerable advantage, especially to the small dairy operation. Yet, as mentioned earlier, every consideration should be given to building in flexibility and versatility even for short-term programs. Considerable advantage may accrue to dairies considering modular construction to meet current needs with an eye to the future.

Cost to Treat

Of greatest impact in process selection is the cost to treat which includes
both capital investment and operational maintenance costs. All too often,
process selection based on first cost has proved to be the poorest choice
owing to excessively high operation and maintenance costs. Considerable
care must be observed in interpreting cost data in the literature. Hidden
costs such as sludge handling and whey separation and treatment are often
neglected in these analyses. Data on municipal treatment process costs
are of little value in assessing costs to treat dairy wastewaters.

TREATMENT ALTERNATIVES

In the discussion of alternatives in dairy wastewater treatment, no reference
will be made at this point to whey treatment or handling. The processing
of whey is extremely important in the overall waste disposal program and
should be considered separately from other wastewater disposal problems.
In the overwhelming majority of cases whey should not be treated in combination with other dairy wastewaters. The characteristics of whey normally
lead to serious treatment plant upsets and would result in excessively high
costs to treat by most procedures currently employed.

Dairy plant wastewaters are most often treated by biological treatment
processes owing to the relatively high fraction of readily biodegradable
compounds present. Since the major fraction of pollutants in dairy wastewaters is dissolved and colloidal organic and inorganic matter, chemical
coagulation and precipitation of milk waste constituents has met with only
partial success resulting in relatively poor removal of organic matter and
producing voluminous quantities of chemical sludges.

Biological processes, then, provide the most economical process for removal
of the substances. The rate of biochemical stabilization of the organic compounds in dairy food wastewaters is normally dictated by the rate of degradation of milk proteins. Limitations of essential growth nutrients such as
nitrogen or phosphorus, the toxic effect of detergents or sanitizers, or inhibition or repression of the activity of specific enzymes caused by lactic
acid or whey proteins may further exert a controlling influence on stabilization rates.

In general, however, a biological process can be designed that will effectively and consistently stabilize a major fraction of the degradable waste
constituents. This may be most effectively accomplished by an assortment
of processes including aerobic systems, anaerobic systems, or combinations
of the two processes in either suspended or fixed film reactors.

The design of biological wastewater treatment processes is dependent upon
(a) the stoichiometry of the biochemical reaction, (b) the rate at which this
reaction proceeds, and (c) the dispersion of the waste constituents within
the reaction vessel. A vast literature exists in sanitary engineering related
to the modeling of biological treatment processes. Investigators recognize
the great complexity of the system and considerable effort has been recently
exerted to develop a unified concept acceptable to the profession. Suffice
it to say that such an agreement is still a long way off.

The biochemistry of milk wastewater stabilization has been the subject of
numerous investigations since the early 1950's. Considerable effort has
provided a general understanding of the mechanisms of milk decomposition
and some general stoichiometric relationships have been reported.

The selection of biological reactor type and the mode of microorganism-
wastewater contact are critical to the expected performance of the treatment system. Normally, the microorganisms may be held in suspension by
aeration or mechanical mixing (suspended growth reactor) or they may be
grown on surfaces over which wastewater is directed (fixed film reactors).
The hydraulic regime produced in either reactor type will also dictate the
apparent performance of the system. Thus, flow configurations described
as plug flow, completely mixed, partially mixed with longitudinal dispersion, batch operation and complete-mixed-in-series are often used.

In practice, the design of biological wastewater treatment processes is
often based upon empirical design parameters and rules of thumb. The
magnitude of these parameters or standards of design are applicable primarily
to domestic wastewater systems and should not be employed for design of
dairy plant wastewaters. A fundamental approach to biological wastewater
modelling in design for industrial wastewaters requires some experimental
study. Where funds are limited, reliance upon information in the literature
is second best, but extreme care must be taken to insure that treatability
of the wastewater in question is realistically parallel to that being used as
the model.

The following sections briefly describe a number of biological processes
which have been employed successfully to treat dairy wastewaters. The
performance of these processes, based upon past experience, is also presented merely to provide some idea of the range of performances expected
under the design conditions. The strengths and weakness of each process
for dairy wastewater treatment application are also cited.

Activated Sludge Process

One of the most popular methods employed for the treatment of dairy waste-

waters is the activated sludge process. Flowsheets of several of the activated sludge process modifications appear in Figures 64, 65 and 66. The
process provides aerobic biological treatment employing suspended growths
of bacterial floc. High concentrations of organisms, maintained by sludge
return, reduce overall reactor size. The organisms are separated from the
treated effluent by means of plain sedimentation.

FIGURE 64:   CONVENTIONAL ACTIVATED SLUDGE FLOW DIAGRAM

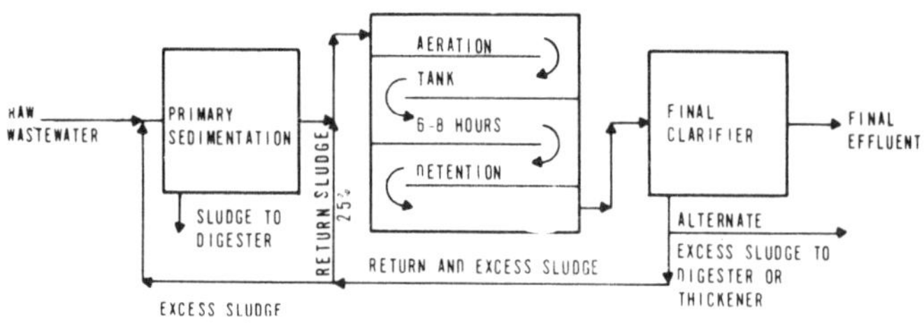

FIGURE 65:   CONTACT STABILIZATION

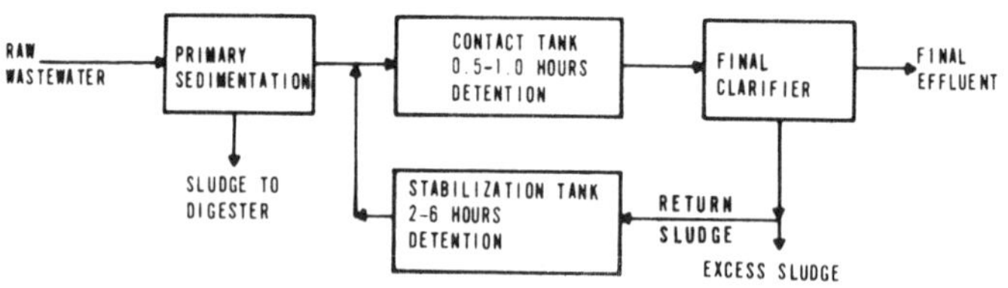

FIGURE 66:   COMPLETELY-MIXED FLOW DIAGRAM

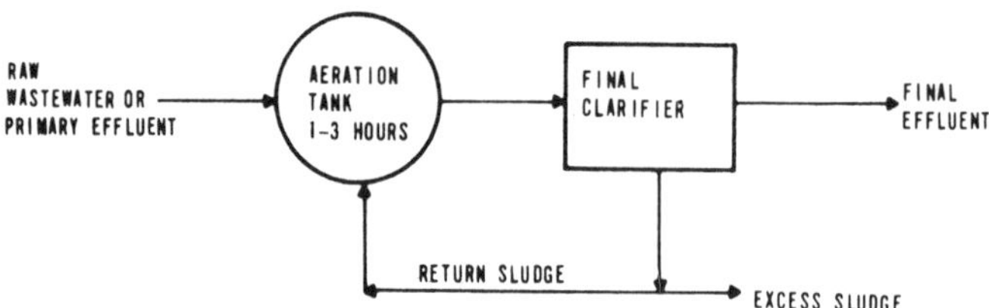

Source: Process Design Manual for Upgrading Existing Plants, EPA
Cont. No. 14-12-933

Dairy wastewaters are normally treated in modifications of the conventional process flowsheet. Experience over the years has indicated that long hydraulic detention times, on the order of 15 to 40 hours, produce most satisfactory process results. In addition, a completely mixed flow regime appears most satisfactory providing an adequate dampening effect on the wide fluctuations in wastewater flows and strengths. The use of long detention time, completely stirred reactors (extended aeration, Figure 66) precludes the use of special flow equalization facilities. Even pH fluctuations normally found in the raw wastewater stream will be significantly attenuated in this system.

Batch fill-and-draw activated sludge systems are also quite popular in the treatment of dairy wastewaters. These processes, too, normally operate over long periods of detention time and provide substantial dampening of flow and strength variations. A very complete compilation of activated sludge performance for dairy wastewaters appears in the Ohio State Report (2). A brief summarization of selected findings from both recent reports on dairy wastewaters (2, 59) are presented in Table 65, along with ranges of design parameters.

It is apparent from examining Table 65 that a wide range of both influent and effluent BOD values is reported over a considerable range of detention times and volumetric loads. In general, superior performances are achieved at longer detention times although conscientious plant operation is paramount to successful performance. Two important design parameters, sludge age [pounds of volatile suspended solids (VSS) under aeration divided by pounds of VSS wasted or lost per day] and loading velocity (pounds of BOD applied per pound of VSS under aeration), are missing from this table. It is difficult to obtain reliable data on these two parameters from the literature, yet they truly define process performance.

The active microbial population (often expressed as VSS in activated sludge systems) is an important variable, too, in describing overall system response. Absence of these values in Table 65 makes a clear delineation of expected performance a difficult task.

Based on past experiences, it appears that BOD removal efficiencies for dairy wastewaters in excess of 90% can be consistently maintained in extended aeration activated sludge plants. More germane, however, is that effluent BOD concentrations below 30 mg./liter may be difficult to achieve consistently, even at long detention periods and sludge ages. Note that for raw wastewater BOD concentrations of 1,500 mg./liter, 98% BOD must be achievable. Of great importance in evaluating the consistent performance of the activated sludge process is the settling properties of the mixed liquor. Bulky sludge is not uncommon in plants treating dairy wastewaters,

TABLE 65: PERFORMANCE OF BIOLOGICAL TREATMENT SYSTEMS — DAIRY FOOD WASTEWATERS

Type	Influent BOD (mg/l)	Effluent BOD (mg/l)	Removal of BOD (%)	Volumetric Load ($\frac{LB\ BOD}{1000\ cuft}$)	Hydraulic Detention (Hours)	Number of Plants Reported
Selected Values Activated sludge	620–1620	13–290	64–99	25–130	15–50	7
Activated Sludge	--	--	24–99.6	---	---	100
Oxidation Ditches	410–2150	3–165	74–99	4–41	27–320	10
Aerated Lagoons	1000	20	98	---	---	1
Aerated Lagoons	---	--	70–99	---	---	5

Performance of Anaerobic Treatment Processes
on Dairy Wastewaters

Type	Influent BOD	Effluent BOD	% Removal BOD	Detention Time	Number Surveyed
Septic Tanks and Digesters	7000–500	300–400	50–87	3–10	12

Source: EPA Technology Transfer Seminar, March, 1973

and the discharge of poorly settled sludge with the effluent will substantially elevate effluent BOD values. Even with extended aeration, as much
as 30% of the effluent volatile solids may contribute to the total effluent
BOD. Oxygen requirements for stabilization of organic wastewaters in
activated sludge are proportional to both the active biomass and the applied
BOD removed. Laboratory or field analyses are employed to ascertain these
requirements.

Oxygen for the activated sludge process may be provided by one of numerous types of diffused air or mechanical aeration systems. Oxygen transfer
rates are dependent upon the aerator selected, the geometrical configuration of the reactor and the wastewater characteristics. This latter information is normally obtained through experimental studies. Diffused air requirements often quoted for dairy wastewaters are usually in excess of 1,500
cubic feet per pound of BOD removed. Coarse bubble diffusion devices
are commonly recommended for dairy wastewaters as they clog less frequently
and require less maintanance than the fine bubble systems. Jet aerators,
shear devices, surface turbines, pumps, draft-tube aerators, rotors and
brushes are also popular. Oxygen transfer rates for these mechanical devices normally range from 3 to 4.5 pounds of standard oxygen per gross
horsepower hour at standard conditions.

The production and handling of sludge from activated sludge plants treating
dairy wastewaters are not well documented in the literature. Long aeration times (extended aeration) are recommended to destroy (endogenous
respiration) a substantial portion of the sludge solids. Nonetheless, provision must be made to handle some sludge since a certain fraction of the
sludge is nonbiodegradable and will eventually accumulate. Disposal of
accumulated sludges from dairy wastewater activated sludges continues to
be a problem as sludge handling may be very costly. Aerobic digestion of
accumulated sludges followed by land disposal represents a desirable relatively low cost alternative.

The activated sludge process is less sensitive to temperature than other biological processes. Toxic effects of sanitizers and pH variations are usually
effectively reduced through the use of extended aeration, completely
mixed systems. Furthermore, the stability of activated sludge settling properties is more effectively enhanced in completely mixed systems employing
low BOD loading velocities.

Finally, long-term aeration employing sludge ages in excess of 10 days will
normally produce highly nitrified effluents (a property likely to be desirable
in the future if effluent standards are adopted for ammonia discharges). The
removal of phosphorus by activated sludge systems is poor, normally ranging
from 35 to 50%. Chemical precipitation, usually prior to final sedimentation,

TABLE 66:　COST COMPARISONS FOR WASTEWATER TREATMENT – BASE YEAR 1963+

| | Capital Costs – $/1000 gal* | | | | |
	Creamery	Cheese	Condensed Milk	Ice Cream	Milk & Cottage Cheese
Ridge & Furrow	325	366	380	375	300
Spray Irrigation	927	1070	1070	1080	850
Aerated Lagoon	540	1170	100	78	1410
Trickling Filter	2030	5050	1350	5050	4050
Activated Sludge	1360	3360	900	3380	2700
LBS BOD/d	153	66	236	2.9	502
Gallons/d	17,200	4100	162,000	6400	19,600
Production–Lb/d	3,900	3400	46,200	890 gal/d	39,500 milk 630 cot.chs.

Operation & Maintenance**

| | Creamery | | Cheese | | Condensed milk | | Ice Cream | | Milk & Cot.Cheese | |
	$/1000gal	$/lb	$/1000gal	$/lb	$/1000gal	$/lb	$/1000gal	$/lb	$/1000gal	$/lb
Ridge & Furrow	0.18	0.02	0.20	0.01	0.21	0.14	0.21	0.47	0.17	0.01
Spray Irrigation	0.51	0.06	0.60	0.04	0.59	0.40	0.60	1.32	0.46	0.02
Aerated Lagoon	0.30	0.03	0.66	0.04	0.06	0.04	0.04	0.09	0.77	0.03
Trickling Filter	1.12	0.13	2.80	0.17	0.75	0.51	2.78	6.10	2.23	0.09
Activated Sludge	0.75	0.08	1.87	0.12	0.50	0.34	1.84	4.06	1.48	0.06

+ From "The Cost of Clean Water – Volume III, Industrial Waste Profile 9, Dairies, FWPCA, Washington,D.C. June 1967 (1)
* Cost per 1000 gal design flow
** Cost per 1000 gal total waste flow or per total pounds of BOD

Source:　EPA Technology Transfer Seminar, March, 1973

employing aluminum or iron salts will effectively remove phosphorus from final effluents and may serve to enhance settling properties of poorly settling sludge.

Cost of the activated sludge process are summarized in Table 66 based on data available in the Cost of Clean Water Series – Volume III, Profile 9 (3). All costs are reported as 1963 dollars, but are the best available data as quoted in this 1973 EPA Report (58). Unless otherwise noted, costs are based on a "medium" plant size with current technology. Capital costs are based on dollars per 1,000 gallons of design flow, whereas operation and maintenance costs are based on pounds of BOD, and 1,000 gallons of wastewater treated.

Considerable caution should be exercised in placing emphasis on these figures since construction costs continue to rise rapidly and local conditions will fluctuate considerably from the national norm. In addition, the level of treatment efficiency required has not been stipulated in the compilation of these figures, but stringent effluent criteria may substantially elevate this figure. Finally, sludge disposal costs are often neglected in these figures, a fact which may lead to serious underestimates of true wastewater treatment costs for those processes generating high sludge volumes.

A process developed by Y. Sekikawa and I. Tanaka; U.S. Patent 3,444,076; May 13, 1969; assigned to Kurita Industrial Co., Ltd., Japan is a process for treating organic wastewater, such as dairy wastewater, in which a mixture of organic wastewater and recycled activated sludge is aerated under superatmospheric pressure, following which the activated sludge is separated and is recycled to the beginning of the process.

Conventional methods for processing organic wastewater employing active sludge, utilize the so-called aerobic treatment wherein the wastewater is mixed with an active sludge comprised of colonies of microorganisms and the resulting mixture is then subjected to aeration.

More specifically, conventional methods of treating wastewater by the use of active sludge or the so-called active sludge methods include several modifications, such as aerating the mixture of wastewater and activated sludge and then separating and recycling the activated sludge, or performing the aeration in several steps, or subjecting only the activated sludge, separated from the aerated liquid, to reaeration and further mixing the reaerated sludge with the wastewater and subjecting the mixture to aeration again. In order to efficiently treat organic wastewater containing high concentrations of organic substances by means of the aforestated active sludge methods, it has been necessary that the wastewater during the aeration step have a high active sludge concentration and that a sufficient amount of oxygen be supplied thereto. However, in the aforesaid conventional methods, both the aeration of the mixture of wastewater and activated sludge and the re-aeration of the activated sludge separated from the aerated liquid are performed under atmospheric pressure. As such, in the case where the wastewater had a high BOD value, it is necessary to continue aeration for an extended period of time. When the BOD value of the wastewater is several thousand ppm, the aeration requires one or more days to complete.

In the conventional active sludge methods, it has been the usual practice to conduct aeration by maintaining the concentration of the active sludge in wastewater in the order ranging from 3,000 to 4,000 ppm. Any further attempt of elevating the active sludge concentration in treating wastewater results in a poor separation of active sludge from the aerated liquid in the subsequent step, making the operation impractical. For this reason, conventional methods for treating such organic wastewater having a very high concentration of organic substances, such as with a BOD of several thousand ppm, uses the technique of diluting the wastewater with a large quantity of pure water before aeration.

Also, the wastewater treatment methods of the prior art which perform aeration under atmospheric pressure, generally separate activated sludge from the aerated liquid by means of a sedimentation technique, and as a consequence, it is quite difficult to maintain the concentration of the active sludge, which is separated in the separation step, at 1% or more. It is, therefore, impossible to elevate the concentration of active sludge in wastewater, even by returning a large quantity of sludge, which has been separated by sedimentation, to the wastewater to be mixed with the wastewater before being subjected to aeration.

As such, the wastewater treating methods of the prior art which conduct the operation under atmospheric pressure are inadequate for treating wastewater having a high concentration of organic substances, and they are, therefore, not efficient methods. Figure 67 shows one version of the process wherein the entire liquid mixture of wastewater and active sludge is maintained under pressure while being aerated.

As shown there, organic wastewater supplied from a raw feed pipe (1) is mixed with active sludge introduced from sludge recycling pipe (2), and the mixture is transferred to a pressure aeration tank (4) by means of a pressurizing pump (3). An air distributing unit (5) is provided in the lower portion of the tank (4). The air distributing unit (5) communicates with an air compressor (6). Air which is compressed by the air compressor (6) is forced into the liquid mixture contained in the tank (4) to aerate the wastewater mixture therein. The pressure within the pressure aeration tank may be in the range of from 2 to 5 kg./cm.^{2} gauge and preferably in the order of the gauge pressure of 3 kg./cm.^{2}. The excessive air which is located in the upper

FIGURE 67:   FLOW DIAGRAM OF PRESSURIZED ACTIVATED SLUDGE
PROCESS

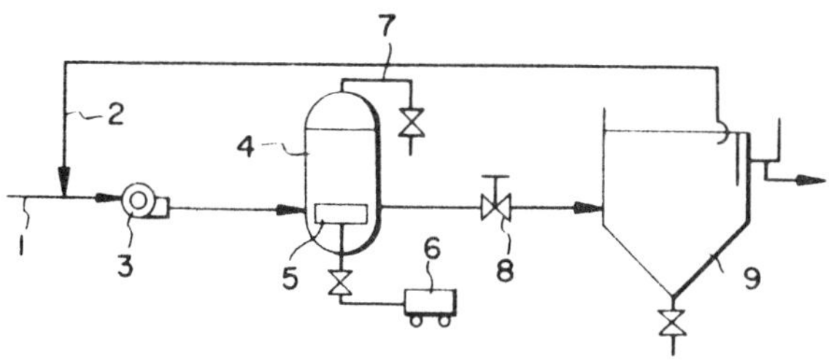

Source: Y. Sekikawa and I. Tanaka; U.S. Patent 3,444,076; May 13,
1969

portion of the tank is gradually discharged outside the tank through pipe (7)
while maintaining the aforesaid pressure in the tank. However, the air may
be partially recycled to the air compressor to save the driving force used
for the aeration. The duration of aeration can vary depending on the degree
of concentration of the wastewater to be treated and also on the volume of
the active sludge to be mixed therewith. However, in view of the fact
that the volume of oxygen dissolving in the wastewater increases substantially because aeration is performed under pressure, the velocity of treatment is accordingly more than doubled as compared to that of the conventional methods, and as a consequence, the length of time required for aeration can be reduced to about 1/2 or less of that of the prior methods.

The method of aeration may employ, other than the aforesaid forcing of compressed air into the aeration tank, for example, a procedure comprising the
spraying of an aqueous mixture of wastewater and active sludge into a zone
of pressurized air. While it is preferred that wastewater be introduced into
the pressurized aeration tank after the wastewater has been mixed with active sludge, the wastewater and the active sludge may be separately introduced into the tank for being subsequently mixed together and aerated within the tank.

The liquid mixture after completion of aeration is introduced through a reducing valve (8) into a separation tank (9) where the liquid mixture is
placed under atmospheric pressure. Whereupon, the air, which until then
has been dissolved in the liquid, is rendered to a colloidal state, comprising
fine bubbles which rise to the portion near the surface of the liquid and remain there floating. As a consequence, active sludge in the tank (9) is
easily separated within a short period of time.

Furthermore, the separated sludge has a concentration of about 3% which is 2.5 to 3 times as much as that obtained by the prior sludge separation method using the sedimentation technique. As such, separated dense active sludge can be recycled to the initial aeration zone, and because of this the active sludge concentration of the wastewater located in the aeration tank can still be maintained at a high level even when the quantity of the recycled active sludge is reduced below the quantity required in the prior art. Such an elevated concentration of active sludge in the wastewater contributes to a further increase in the speed of aeration, and this enables the wastewater having a high concentration of organic substances to be treated within much less time than that required in the prior methods.

Since, in the process, aeration is conducted under pressure, as has been described above, a large amount of oxygen can be dissolved in the wastewater, and as a consequence, treatment velocity can be more than doubled over that in the prior art wherein aeration is conducted under atmospheric pressure. This permits aeration to be performed in a reduced amount of time without requiring dilution of highly concentrated organic wastewater prior to aeration, and also permits the use of an aeration tank with a reduced capacity.

Furthermore, activated sludge is allowed to separate from the liquid and to rise to the surface of the liquid within a short period of time by simply placing the aerated liquid under atmospheric pressure, and still the separated sludge is of a greater concentration than that obtained by the prior art. Therefore, highly concentrated active sludge can be recycled so that the concentration of active sludge in the wastewater is elevated and accordingly the operation velocity can be further increased. The following is a special example of the operation of the above apparatus.

Into the wastewater having a BOD value of 4,130 ppm from a dairy product factory was mixed active sludge having a concentration of 2.7% and this mixture was subjected to aeration, with the concentration of the sludge in the aeration tank being maintained at 13,100 ppm and the pressure being held under a gauge pressure of 3 kg./cm.^{2}. The volume of air supply per minute was identical with the volume of the aerated liquid supplied (converted as the volume under atmospheric pressure). After 4 hours, the BOD_5 of the treated water was noted to be 135 ppm. The aerated liquid was separated from the sludge in 15 minutes by the flotation method and the concentration of the separated active sludge was 2.7%.

The same type of wastewater was subjected to aeration under atmospheric pressure while maintaining the active sludge concentration in the aeration tank at 14,000 ppm and supplying air in a volume identical to the volume which has been described above. The BOD_5 of the treated water after a treatment of 4 hours' duration was noted to be 1,020 ppm, and it required

9 additional hours of treatment to reduce it to 120 ppm. The resulting
aerated liquid was separated by sedimentation, and separation was completed in 1 hour. The separated active sludge was noted to have a concentration of 1.6%. In this instance, the activated sludge was condensed
by a centrifugal separating machine before it was mixed into the wastewater in order to control the concentration of the active sludge which was
initially supplied to the aeration tank.

Oxidation Ditches

The treatment of milk wastewaters in oxidation ditches has been acceptable
practice for a number of years in Europe. The oxidation ditch is an extension of the activated sludge process employing a ring-shaped circuit or ditch
usually 6 to 10 feet deep (Figure 68). Aeration is provided by cage or brush
aerators mounted at several points along the ditch in order to circulate flow
around the circuit and to maintain sufficient velocity to keep solids in suspension. Baffling of rectangular lagoons with appropriately located flow
directors will achieve the same effect. The oxidation ditch may operate as
a continuous system or as a batch process. If operated in a continuous mode,
a clarifier is incorporated as an integral part of the system.

FIGURE 68:   OXIDATION DITCH

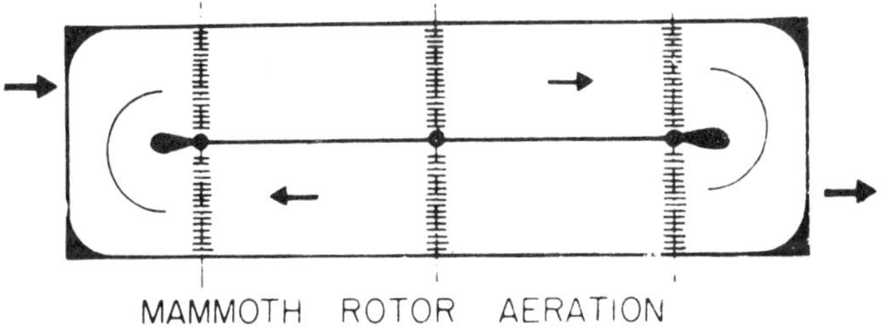

Source: EPA Technology Transfer Seminar, March   1973

Typical performance data for oxidation ditches appear in Table 65. As
was noted with the activated sludge process, considerable variation in
loading, detention time, and process efficiencies are apparent. There is
considerable evidence in the literature that at detention times in excess of
50 days and at volumetric loadings less than 15 pounds BOD/1,000 ft.^{3}
effluent concentrations of less than 30 mg./liter are achievable. High
mixed liquor VSS, in excess of 4,000 mg./liter, are attainable with the
configuration when employing sludge recycle. There is no evidence to
suggest that this particular configuration will more successfully treat dairy

wastewaters as compared with the extended aeration activated sludge
processes at similar loadings. Costs of this process are likely similar to
those for activated sludge although there is not enough operating experience
in this country to provide reliable cost data.

Aerated Lagoons

Aerated lagoons are also an extension of the activated sludge process wherein no sludge return is normally practiced. As a result, the active biomass
(VSS) in the lagoon is low, thereby requiring longer periods of aeration for
comparable performance.

Data in the literature on the performance of aerated lagoon systems for dairy
wastewaters are sketchy. Only 3 reports appear in the comprehensive survey
by the Ohio State group (2), and 5 are reported by A.T. Kearney & Co.
(59); (Table 65). From this data there is evidence that current effluent requirements can be met by properly designed aerated lagoon systems. Design would most definitely be predicated upon careful pilot or laboratory
scale studies. The experience of these authors is that aerated lagoons are
a satisfactory alternative for dairy wastewater treatment. Currently a number of these systems for dairies are in successful operation in Wisconsin.
An example of two of these systems will be presented later.

In most cases aerated lagoons are not vigorously mixed, resulting in sedimentation of suspended solids within the lagoon itself. These "facilitative"
aerated lagoons, therefore, remove organic matter through physical separation and anaerobic and aerobic stabilization. Mixing intensities normally
required to prevent sedimentation in aerated lagoons require power inputs
of approximately one order of magnitude greater than for oxygen dispersion
alone (8 HP/MG vs 80 HP/MG).

On the other hand, horsepower requirements for aeration are dependent
upon the oxygen uptake rates, the required hydraulic detention time, and
the oxygen transfer characteristics of the wastewater. In most instances
the power required for aeration is less than that required for complete solids
suspension and the engineer must determine whether the added power costs
justify this added mixing capacity. In the majority of cases, dairies have
not opted for this extra expense, a decision which appears sensible. Aerated
lagoons are normally designed as a series system. Most aeration is provided
in the first cell, followed by a second or third cell of quiescent settling.
Normally lagoon depths of 10 to 15 ft. are desirable although shallower cells
are allowable for settling. Algal growths occurring in the quiescent cells
may cause a deterioration in effluent quality and should be avoided through
proper outlet design, covering or filtration. Aerated lagoons are temperature sensitive, producing poorer quality effluents in the winter months. Most
engineers size aerated lagoons based on winter operations. The onset of

warmer temperatures in the spring may result in increased biological activity
in the anaerobic zones of facultative aerated lagoons. This activity often
results in depletion of lagoon dissolved oxygen causing odors and loss of
efficiency. Aerator designs should provide for this eventuality, especially
in the poorly mixed lagoon systems.

Aeration is normally provided by either surface or submerged aeration equipment. In northern climates, surface aerators require considerable maintenance for ice removal and in retaining effective and consistent operation. Submerged units will not normally be affected by cold weather but
orifices may clog (as with activated sludge) and mixing velocities are usually low.

Land requirements for aerated lagoons are high and most states require substantial distances be maintained between lagoons and residences. Because
of long detention times, no flow equalization is required when employing
lagoon systems. Sludge handling is relatively minor in most lagoon systems,
although some provision may be made to dewater quiescent lagoons in the
event of significant solid buildup. Anaerobic digestion normally maintains
a relatively small sludge volume on the lagoon bottom.

Nitrogen conversion to nitrate is usually complete in aerated lagoons.
Phosphorus removal is reported to range from 30 to 80% depending upon the
season of the year. Phosphorus precipitation would account for this removal
and resolubilization of precipitated phosphorus is likely to occur during
certain periods of the year. Costs for lagoon construction and operation
are presented in Table 66. Land costs were estimated at $300.00 per
acre (3).

Stabilization Ponds

Stabilization ponds cover a variety of lagoon systems employed for wastewater treatment. As compared with aerated lagoons, stabilization ponds
depend upon surface reaeration and photosynthesis for oxygen supply. For
dairy wastewaters with strengths in excess of municipal wastewaters (300 mg.
per liter BOD), lagoon surface areas or active algal populations must be extremely large. There is no practical method currently available in the midwest for maintaining algal cultures in the concentrations necessary to effectively treat most dairy wastewaters. Solar insolations are too low and winter
conditions too severe for successful operation.

Surface area requirements would appear prohibitive for most dairy wastewaters, being approximately 15 to 20 lbs. BOD/acre/day. Thus, a dairy
producing 500 lbs. BOD/day would require at least 25 acres of land for
stabilization ponds. In addition, at least 1/4 mile must be maintained
between lagoon and the nearest residence.

The Ohio State Report (2) represents a compilation of performance of stabilization ponds. With few exceptions, effluent BOD exceeds that currently required by current effluent quality standards. The State of Wisconsin has stated that, if the New Federal Water Pollution Control Act is enforced as written, "treatment processes such as....stabilization ponds would no longer be permitted as the sole means of treatment." Based on the factors discussed above, there would appear to be little advantage in considering this method of treatment for dairy wastewaters.

Trickling Filters

The trickling filter process in contrast to suspended growth processes employs a fixed support medium to maintain the active organisms within the wastewater stream. In the past this medium has been rock, slag, or other low cost material providing a large surface area per unit volume with a high void volume. Recently, numerous types of low weight, high specific plastic media have been developed for this purpose.

Organic matter associated with the wastewater is absorbed or adsorbed into the fixed biological film and is subsequently oxidized. Oxygen is normally provided by natural ventilation within the fixed bed although positive airflow may be provided to achieve more effective process operation. Contact time of the waste in trickling filters is normally short depending upon the application rate to the filter and the filter depth. Conventional filters are 6 to 8 feet deep whereas the plastic media filters may be constructed as high as 30 feet. Increased contact time is provided in some filters by recirculation of treated wastewater back through the filter. The mode of recirculation varies considerably from plant to plant.

Trickling filters are followed by clarification facilities in order to intercept and remove sloughed solids from the filter. Filter sloughing is a natural process and must occur in order to maintain an active biological mass on the media. Heavy accumulation of biological growth on the media surface will lead to filter clogging and poor oxygen transport.

There are numerous flowsheets currently employed that use the trickling filter process. The engineer selects a flowsheet which makes best use of the existing site and provides greatest operational flexibility for the wastewater being treated. Filters are normally designed based upon either hydraulic load (millions of gallons per acre per day MGAD) and organic load (pounds of BOD per 1,000 ft.^{3} per day) and they are normally classified as high or low rate in accordance with these loading parameters. Considerable controversy still surrounds the selection of the appropriate design parameter and its order of magnitude.

The performance of trickling filters treating dairy wastewaters has been

summarized in the Ohio State Report (2). Examination of this extensive
tabulation is confusing and of little real value to design engineers. As
with the activated sludge tabulations, wide variation exists in both performance and magnitude of design parameters. Scatter diagrams prepared
in the report (2) would suggest that neither hydraulic load nor organic load
control process efficiency to any great extent. Process efficiencies range
from less than 10% to 99% over ranges of hydraulic loading of from 0.14 to
20 MGAD and organic loads ranging from approximately 2 to 175 lbs. BOD
per 1,000 ft.3/day. Clearly, many of these reports are for roughing filters
employed for pretreatment only. The A.T. Kearny & Co. Report (59) summarizes performance efficiencies ranging from 35 to 99.8% for 48 plants reported.

It is clear from this data that trickling filters, properly designed, may
achieve current effluent quality standards. Indications are that plants designed at hydraulic loads less than 2.0 MGAD and organic loads less than
20 lbs. BOD/1,000 ft.3/day may have a reasonable likelihood of success
in achieving low effluent BOD. Such generalizations, however, are not
sufficient to be employed by dairy food processors without considerable
investigation. Currently, the State of Wisconsin has not favored trickling
filtration as an effective means of secondary treatment. Winter operation often deteriorates effluent quality and covering of existing filters is
strongly recommended in northern climates.

The proper ventilation or oxygen transfer in trickling filters is paramount to
to successful performance. Dairy wastewaters exert high oxygen demands
as compared to municipal wastewater (per unit volume of waste) thereby
putting a high demand on oxygen resources in the filter. Poor air circulation caused by heavy biological growths, clogged underdrains, and waste
channeling will result in serious odor conditions, development of massive
biological growths and deterioration in effluent quality. Positive ventilation procedures may effectively be employed to improve operation.

For this reason it is felt by these authors (59) that the controlling design
parameter in the case of high strength wastes is organic loading. Recirculation of wastewater through the filter is often advantageous to eliminate
excessive growths, reduce filter fly populations, and reduce odor. Cooling effects of recirculation are undesirable, however, during the cold winter months. The use of plastic media filters for dairy wastewater treatment
may prove to be most advantageous. More uniform void volumes and high
specific surfaces will promote more effective oxygen transfer and wastewater contact with the biomass. In addition, considerable savings may be
realized in land area when deep filters are employed.

Trickling filters handle shock loads moderately well although effluent quality may suffer for short periods. Whey should never be applied unless slowly

added over long periods of time and filter design should account for this
addition. Nitrogen conversion to nitrates will occur only on lightly loaded
filters (usually less than 5 lbs. BOD/1,000 ft.^{3}/day) and phosphorus removals are poor, usually being less than 35%. Sludge produced in trickling
filters, although less voluminous than that from activated sludge processes,
must be subsequently handled. Anaerobic or aerobic digestion of sludges
followed by land disposal are most commonly applied procedures.

The cost of trickling filter treatment appears in Table 66. Plastic media
filtration may be more expensive than indicated in this tabulation; however,
a higher quality effluent will normally result more consistently than with
conventional rock filters.

Rotating Biological Discs

The rotating disc process is a modification of the trickling filter process
whereby the fixed biological film is rotated through the wastewater. First
developed in Germany in 1955, there are now over 1,000 installations in
Europe alone. Research in the United States has developed a lighter, lower
cost disc providing higher surface area. A large biological surface is provided by a series of closely spaced discs mounted on a rotating horizontal
shaft. (Figure 69). The discs are slowly rotated at about 2 rpm through the
wastewater while submerged to about 40% of their area. Organic matter
is sorbed by the biomass on the disc and is subsequently oxidized in the
presence of oxygen. A positive means of excess film sloughing is provided
by the shearing action caused by the rotation of the discs.

The biological discs are normally staged so that a number of discs rotate
within a given enclosed reactor cell. Wastewater passes from cell to cell
through openings in the cell walls. This separation of reactors in series
provides an advantageous development of specialized biological cultures
for the waste constituents during each phase of treatment. Thus, by adding
additional cells, one may achieve progressively higher levels of oxidation
including complete nitrification of the wastewater. This cellular structure
also reduces the effect of shock loads to the system. A clarification facility is required to remove sloughed biological solids from the discs.

Performance data on the biological disc in treating dairy wastewaters are
scant. An example of one such application is given later in this paper.
Design parameters currently used for the process are based upon hydraulic
loading (gallons per day per ft.^{2}) and plant staging. Sludge production in
these units is normally comparable to that in trickling filters and methods
of handling and disposal parallel those discussed earlier. Costs of this type
of treatment are still preliminary although it is apparent that operation and
maintenance costs will be very low. The only power consumed is that used
to rotate the disc shaft.

FIGURE 69:   ROTATING BIOLOGICAL FILTER

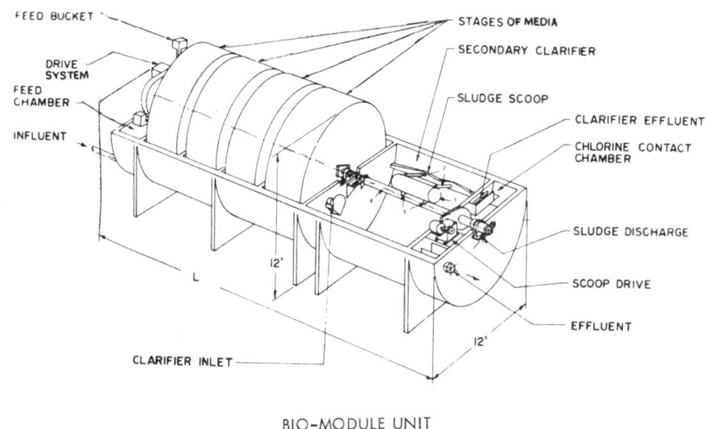

BIO-MODULE UNIT

Source: EPA Technology Transfer Seminar, March, 1973

Anaerobic Processes

The anaerobic treatment of dairy process wastewaters has been practiced
for many years in small dairy operations through the use of septic tanks.
During anaerobic decomposition, lactose is rapidly converted to lactic
acid, lowering the pH. In addition, fats and proteins are decomposed to
amino acids, organic acids, aldehydes, alcohols, and other anaerobic intermediates. This phase of biodegradation is often referred to as acid fermentation and little BOD, COD or organic carbon "removal" is achieved.

A second phase of biochemical reactions, methane fermentation, may also
proceed, converting the organic acids to methane and CO_2. This gasifica-
tion step subsequently removes BOD from the system as a gas. At steady
state, one reaction feeds the other resulting in a relatively constant pH
and organic acid level. If conditions within the reactor become unfavor-
able for the methane bacteria (the most sensitive of the two groups of bacteria in this reaction system), acid buildup will occur causing further deterioration of the process, a decrease in pH and a reduction in gas production. Successful anaerobic processes, designed for BOD removel, must
develop a successful balance between these two phases.

Anaerobic processes normally have not been successful as complete treatment systems, since effluent quality is often poorer than that required by
stream standards. The process does, however, offer a successful low cost
process for wastewater pretreatment. Even if only the acidification step is

achieved in the anaerobic process, considerably higher rates of aerobic stabilization may be realized with this pretreated effluent. More rapid decomposition of organic acids and anaerobic intermediates under an aerobic environment may result in smaller and more effective aerobic processes than could be achieved without this pretreatment.

Although quiescent holding tanks (septic tanks) have been used to treat dairy wastewaters, improvements in the anaerobic process have resulted in considerably better overall performance. Anaerobic contact processes, employing mixing with sludge return have proved to be successful in accelerating the conversion of organics to methane and CO_2. The use of anaerobic fixed film contractors (anaerobic trickling filters and biological discs) have also been examined.

Results of several reported experiences with anaerobic processes appear in Table 65. Little data is available on anaerobic contact processes and fixed film reactors. The evidence indicates that approximately 50% of the applied BOD can be removed in quiescent digestion systems with little advantage gained beyond 4 days of detention time. Greater removals may be realized through more effective contact between the biomass and wastewater. Imhoff tanks may also provide more consistent results due to the separation of digestion and sedimentation processes.

The anaerobic process is sensitive to shock loads, temperature and toxic chemicals. Process efficiency may seriously suffer when temperatures of the wastewater drop below 10°C. Certain sanitizers, high concentrations of ammonia and numerous metal cations are toxic to methane bacteria. Anaerobic systems must be covered and should be ventilated adequately. Safety precautions should be taken owing to the presence of methane.

Costs of anaerobic systems are not readily available according to Boyle & Polkowski (58). Precast septic tanks range from $400 to $600 including installation up to 1,500 gallon capacity. In most instances use of precast tanks in series is more economical than larger cast-in-place tanks. Operation and maintenance costs for anaerobic processes are dependent upon whether mixing is employed. Costs for simple septic tank operation range from $100 to $200 per year depending upon pumping and hauling costs. Low operation and maintenance costs are due largely to the absence of power requirements and the infrequent need to dispose of accumulated sludges.

Irrigation

The use of irrigation as a treatment and disposal method for dairy plant wastewaters is most efficacious. It represents the best alternative if the proper type of soil is available in large enough acreage. The success of irrigation methods depends upon the use of proper application rates and the effective pretreatment of the wastewater prior to disposal. Careful attention must be paid to the alternating of irrigation plots and provision and maintenance of the cover crop. High sodium concentrations may seal soil particles. Wastes high in particulate matter should be adequately screened prior to irrigation. Recently, considerable success has been achieved by employing the soil as a biological filter. Permeable soils are underlain by tile fields which intercept the percolating wastewater. Nutrients are removed by the soil and its cover crop yielding percolates of high quality. Recycling or additional polishing of the percolate may be required prior to surface discharge.

Ridge and furrow irrigation of dairy wastewaters has met with little success in the upper midwest. Odor nuisance, standing water, and maintenance difficulties are attributed to this method of land disposal. A summary of irrigation practices is detailed in the Ohio State Report (2). Costs for irrigation methods are presented in Table 66. Costs for land are based on a value of $300/acre.

Filtration

The filtration of wastewaters normally provides a polishing step prior to final discharge. Filtration may be provided by microscreens or granular filtration devices employing diatomaceous earth or sand or mixed bed filters of materials such as anthracite and sand or activated carbon and sand. The state of the art of wastewater filtration is relatively new and considerable research continues to improve filtration techniques. Further details may be found in Culp and Culp (60).

Slow sand filters, 12 to 30 inches deep, employ filter sand for approximately 1/2 that depth underlain by coarse sand, gravel, and an underdrain system. Application rates of up to 3 gallons/ft.^{2}/day are employed. Filter cleaning is employed when headlosses reach 8 to 10 feet. The filter is then partially dried and the top layer of sand and sludge are removed. Clean sand is subsequently added prior to placing the filter back in operation. In general, slow sand filters are expensive to operate, require large areas and provide only moderate performance. Rapid sand filters on sand or mixed bed filters are considerably more popular for tertiary treatment processes. Application rates of 3 to 6 gallons/ft.^{2}/minute are commonly employed. Cleaning of filters is provided by backwashing the filter with treated effluent.

In treating dairy wastewaters, the filtration step may be required as a polishing operation to achieve the desired BOD or solids concentrations. If the selected treatment process results in effluent concentrations high in degradable suspended solids (lagooning or even activated sludge processes) filtration may prove feasible. In most cases, filtration processes may be added to most process flowsheets at a future time to improve overall effluent efficiencies. If the treatment process achieves complete stabilization of the wastewater or if suspended solids are mineralized, filtration may not provide much advantage to the overall process flowsheet.

Chemical Methods

Chemical precipitation of dairy food wastewaters is not widely practiced primarily because of its high cost and its nominal effectiveness in organic matter removal. In some cases, wastewaters high in fats or colloidal matter might be effectively pretreated by addition of metal cations such as calcium, aluminum or iron or by polyelectrolyte additions. Voluminous amounts of sludge normally result with the metal salts requiring expensive methods of sludge handling and disposal. Polyelectrolytes may provide advantage heretofor not attainable with other chemicals. Small quantities of polyelectrolytes may affect substantial removal of solids at a reasonable cost. To date, the state of the art is not advanced enough to justify chemical methods as a reasonable alternative in the majority of cases.

A process developed by J. Hollo, J. Toth and I. Zagyvai; U.S. Patent 3,738,933; June 12, 1973; assigned to Tatabanyai Szenbanyak, Hungary is one in which the biochemical oxygen demand of sewage of the type of dairy waste, is reduced by adding to the sewage at least 120 g./m.^{3} of a water-soluble aluminum salt or bivalent or trivalent iron salt, 0.5 to 1 kg. per cubic meter of bentonite or kaolin and 5 to 10 g./m.^{3} of polymers or copolymers of acrylic acid-acrylic amide in the form of an aqueous solution.

The suspension is stirred and the pH adjusted to exceed 10 by addition of a basic calcium compound such as lime milk or calcium hydroxide. The precipitate thus obtained is separated by settling and transferred to a conical tank for treatment with carbon dioxide until the pH is no higher than 7, and the resulting precipitate is filtered and sterilized at a temperature above 130°C. under pressure, to produce a biochemical culture medium or animal feed.

Membrane Processes

The use of molecular sieves, electrodialysis, and reverse osmosis membrane systems has only begun. Considerable success for whey treatment has been predicted. Where water reuse or product recovery is feasible, these methods may prove successful. Costs at this time are high and normally discourage use in most dairy wastewater treatment schemes.

Two broad scope potential applications of desalination machinery exist in the processing of whey from cheese making operations, however (61). These are the primary concentration of whey solids through the use of reverse osmosis equipment; and the demineralization of whey through electromembrane

processing so that the milk sugar and protein have substantially higher value
as a human food substance. The first process, the concentration of whey
from 6% solids to approximately 15% solids by reverse osmosis is now in
the advanced research and pilot stages and has been shown to be practical
but has not yet been put into wide-scale production. The second process,
transport depletion or electrodialysis, is now being industrially carried on
in the United States and the annual production of demineralized whey solids
is approximately 25 million pounds.

The concentration and demineralization of whey solids is an attractive
process not only from the waste handling aspect, but also from the standpoint that in whey there is a substantial volume of high quality protein
which could be utilized in alleviating food shortages throughout the world.
The present uses of whey have only touched upon the relatively high value
specialty markets such as synthetic milk products, special cheese foods,
dietary supplements, infant formulas and substitute low-fat dairy products.
The further researching and application of whey protein and solids to low-
cost, high-volume food products gives rise to the thought that ostensibly
all of the whey production could be utilized in solving food scarcity problems.

The use of reverse osmosis plants for the concentration of whey is a subject
that has not yet been studied in sufficient detail to properly determine the
complete market potential. However, the concentration from 6% solids to
approximately 15% solids without application of high operating pressures
results in the removal of about 2/3 of the water in whey at an operating
cost far below the cost of typical evaporator operations. At this time, it
is felt that the ultimate market potential for reverse osmosis in whey concentration is about 63 million dollars. It is estimated that 10% of this market, or about 6 million dollars will be fulfilled in the 1970-75 period and
that an additional 10 million dollars of reverse osmosis equipment will be
sold in the 1975-1980 period.

Based on presently collected data, it is estimated that the total potential
market for transport depletion and electrodialysis equipment in the United
States cheese producing plants would be about 106 million dollars. During
the 5 years from 1970 to 1975, it could be estimated that approximately 10%
of the market for electromembrane deashing plants might be filled. This
would amount to a total dollar volume of 10 million dollars. During the
second 5 year period from 1975 to 1980, a fair assumption would be that an
additional 15 million dollars of equipment, or about 15% of the ultimate
market may be filled.

New developments combining the technology of electromembrane, reverse
osmosis, and ultrafiltration equipment may lead to additional areas of desalination technology application. Processes now in research stages indicate
that substantial fractionation and upgrading of the protein and lactose com-

ponents of whey can be performed through these methods. The culmination
of this research will result in additional markets for membrane processing
equipment. The direct application of desalination machinery in the cheese
and dairy processing industry is primarily for the treatment of effluents from
cheese plants. Other dairy wastes are primarily spillage and washing from
mechanical equipment and since the product in these cases is contaminated
with various cleaning agents and detergents, recovery is not deemed practical.

Figure 70 shows the estimated value of equipment for reverse osmosis, transport depletion, and electrodialysis devices in the cheese industry. An estimated 1,260 plants of each type (reverse osmosis and electromembrane)
could be used in cheese making operations in the United States.

The transport depletion and electrodialysis equipment used for demineralizing
of cheese whey is generally of the same construction as that utilized in
water desalting operations. The use of sanitary pipes and pumps on the
whey systems increases the prices somewhat over comparable standard water
treatment equipment. Electrodialysis stacks themselves are sold at prices
ranging from $8.00 to $12.00 per square foot of membrane surface depending upon the size of the overall system. Complete operating installations
including pumps, piping, controls, stacks and necessary accessories typically range in price from $15.00 to $25.00 per square foot of membrane
surface. The equipment cost figures used for whey processing plants are
based on proprietary information regarding membrane loading and production rates and on equipment cost figures in the general ranges outlined above.

Current market values for salty, dried whey range from 4 to approximately
8 cents per pound. The lower priced materials are generally considered
animal food grade and are not utilized in food products for human consumption. Purer grades are produced for very limited use in human food products
and sell at prices ranging from 6 to 8 cents per pound. However, the use
of these products is severely limited by the salt levels. Desalted whey
produced by either transport depletion or electrodialysis methods have current market values ranging from 15 to 30 cents per pound. Typical desalted
whey having approximately 1/2 to 2/3 of the salts removed and containing
either a protein enriched balance or a normal protein lactose balance sell
for between 18 and 23 cents per pound.

The cost of operation of an electromembrane process for upgrading whey
solids through demineralization varies greatly, dependent upon the size of
the installation and the amount of labor and overhead costs allocated to
the process. However, the general cost for demineralization of whey
through either transport depletion or electrodialysis techniques is in the
range of 3 to 8 cents per pound of whey solids. To that operating cost figure should be added the amortization of demineralization equipment which

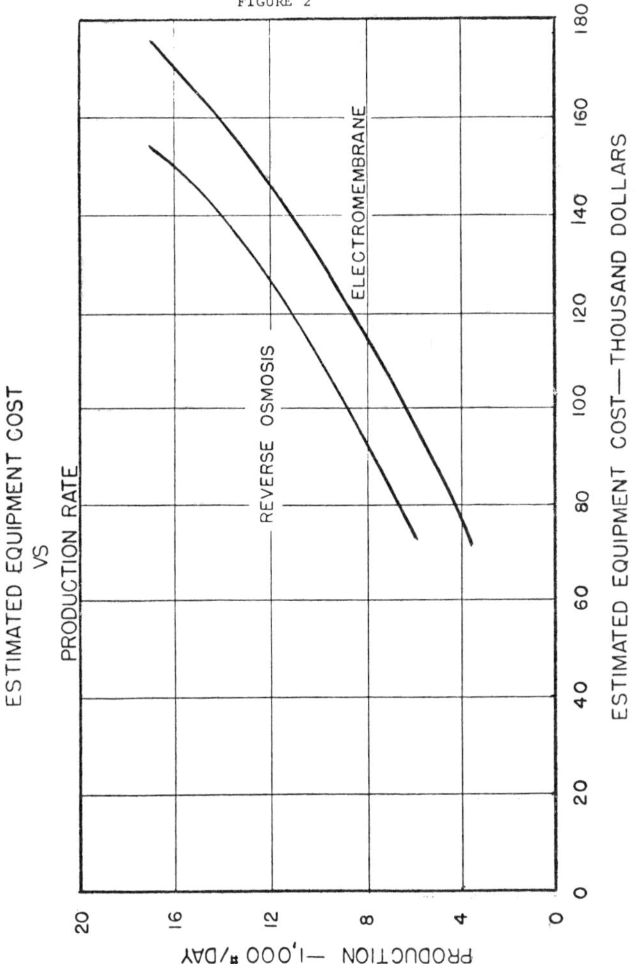

FIGURE 70: DEMINERALIZATION OF WHEY

Source: OSW R & D Progress Report No. 581, October, 1970

would add an additional burden of approximately 2 to 5 cents per pound of whey solids. Reverse osmosis preconcentration of whey would cost approximately 2 to 3 cents per pound while the finishing concentration by conventional dairy evaporators would cost an approximately equal amount. The spray drying of whey powder from liquids of concentration ranges between 35 and 60% solids is an operating cost ranging between 3 and 4 cents per pound. Thus, a high-grade food product worth between 18 and 25 cents per pound can be obtained at a manufacturing cost of between 9 and 19 cents per pound.

Carbon Adsorption

Carbon adsorption technology in the treatment of wastewaters has rapidly advanced over the past 5 years. Carbon adsorption systems are becoming competitive with biological treatment processes in the treatment of municipal wastewaters. Carbon systems are normally sized on the basis of mass of COD removed per mass of carbon. Carbon requirements for dairy wastewaters would be high and the cost of treatment is still not competitive with other alternatives. As effluent requirements increase, however, it is not inconceivable that carbon adsorption will be the competitive choice. A detailed discussion of carbon adsorption theory and application may be found in Culp and Culp (60).

Treatment Methods — Summary

A brief tabulation has been provided to aid a summary evaluation. Table 67 gives some guidance as to each process characteristic. In selecting a treatment process, all alternatives must be considered. The preliminary design may be based to some extent upon values reported in the literature; final selection, however, should be based on treatability tests. This becomes more critical as effluent requirements become more stringent. The data presented in Table 67 relate primarily to the effectiveness of these processes to achieve current effluent standards (30 mg./liter BOD consistently). As effluent standards become more stringent, it will be more difficult to achieve "excellent" performance from many of these processes.

The processes discussed above were considered as separate treatment entities. In practice, it is wise to look at combinations of these processes so that advantages may be taken of the best parts of each. Thus anaerobic pretreatment followed by aerobic polishing may provide a more economical alternative than the aerobic process alone.

Table 68 briefly summarizes the characteristics of a number of selected "tertiary" wastewater treatment or polishing processes, the selection of which is dependent upon local effluent requirements, the current treatment process employed, and the characteristics of the treated wastewater.

TABLE 67:   COMPARISON OF TREATMENT ALTERNATIVES

| Type | Effluent Quality | | | Reliabi-lity | Cost | | Land Reqmt. | Response to Shock | Economic Life (y rs) |
	BOD/Solids	Ammonia	Phosp.		Capital	O & M			
Activated Sludge (Ext.Aer)	+++	+++	+	+++	H	H	L	+++	15
Oxidation Ditch	+++	+++	+	+++	H	H	A	+++	15
Aerated Lagoon	+++	+++	+	++	A	A	H	+++	20
Stabil.Pond	+	++	+	+	L	L	H	++	30
Trickling Filter	++	++	0	++	H	H	A	++	15
Biological Disc	+++	+++	+	+++	H	L	L	+++	15
Anaerobic Processes	+	0	0	+	L	L	L	+	20
Irrigation	+++	+++	+++	++	A	A	H	+++	20

Key:

+++ – Excellent   H – High

++ – Good   A – Average

+ – Fair   L – Low

0 – Poor

Source: EPA Technology Transfer Seminar, March, 1973

TABLE 68:   TERTIARY TREATMENT PROCESSES — EFFECTIVENESS IN EFFLUENT POLISHING

Process	Effectiveness in Removing:				
	Susp.Solids	BOD	COD	Nitrogen	Phosphorus
Microscreening	+	+	0	0	0
Sand Filtration	++	(+)*	0	0	0
Granular Carbon	++	++	++	0	0
Lime Clarification	++	+	0	0	++
Lime Clarification & Dual Bed Filtration	++	+	0	0	++

Key:

++ Excellent $>$ 80%   *Depends on nature of Suspended Solids

 + Good $>$ 50%

 0 Poor $<$ 50%

Source: EPA Technology Transfer Seminar, March, 1973

MUNICIPAL WASTEWATER TREATMENT
AND JOINT TREATMENT

Dairy food plant wastes are treated primarily by biological oxidation
methods, with over 90% of all the dairy food wastewater being treated in
municipal systems. Where the municipal systems are receiving more than
50% of their BOD_5 from milk wastes or more than 10% of the BOD_5 from
whey, the treatment plant takes on characteristics of a dairy food plant
waste treatment system.

Text books and many authors repeatedly state that dairy food plant wastes
are easier to treat than municipal wastes. By this, most mean that because
of the relatively low suspended solids, it is not necessary to go through an
extensive solids removal prior to biological oxidation. However, in practice, it has been shown that the biological oxidation phase of waste treatment is much more difficult than for municipal waste systems, primarily
because of the high organic load and because of the variability in both
hydraulic and organic loading.

DAIRY WASTE COMPATIBILITY IN MUNICIPAL SYSTEMS

Wastewater Characteristics

In comparing dairy wastewaters with domestic sewage, it is apparent from
Table 69 that there are some significant differences that have a bearing on
the treatability and the influence of dairy wastewater on municipal treatment systems. It is apparent that dairy wastewater is a strong waste in
terms of most parameters reported when compared to domestic wastes. The
BOD values are high and vary widely, but the fact that the BOD values are
high is also indicative that the wastewater is amenable to biological treatment. Because the values of BOD for dairy wastes are considerably higher

than that of typical strong domestic wastes, often surcharge rate structures are applied when discharged to municipal systems. The suspended solids or filterable solids are higher than domestic wastewater, but the solids in dairy wastewater are too finely divided to permit separation by gravity settling whereas for domestic wastewater there is a high fraction of suspended solids that are settleable and, thus, amenable to primary sedimentation treatment.

Primary sedimentation practices usually provide good removals at low operating and capital costs relative to the biological treatment costs associated with secondary treatment. It is apparent in Table 69 that the average phosphorus and grease content of dairy wastes are higher than normally expected in domestic wastewater. Removal costs in municipal treatment related to P removal may be assessed to the contributing source thereby increasing overall treatment costs for use of municipal systems. The range of grease concentrations encountered in dairy wastes may exceed acceptable limits imposed by ordinances.

TABLE 69:   COMPARATIVE WASTEWATER CHARACTERISTICS

Wastewater Characteristic	Dairy Wastewaters				Domestic Sewage (9)		
	Harper (6)		Numerow(10)	Other	Strong	Medium	Weak
	Range	Ave.					
BOD, 5 day 20°C	4'50 – 4790	1885	1890	*15-4790	300	200	100
Solids, total	135 – 8500	2397	4516		1200	700	350
Dissolved, Total	---	---	3956		850	500	250
Fixed	---	---	---		525	300	145
Volatile	---	---	---		325	200	105
Suspended, Total	*24 – 5700	---	560		350	200	100
Fixed	---	---	---		75	50	30
Volatile	*17 – 5260	---	---		20	10	5
Settleable Solids ml/liter	---	---	---	0.3-5.0(11)	20	10	5
Nitrogen, Total as N	15 – 180	76	---		85	40	20
Organic	---	---	73.2		35	15	8
Ammonia	---	---	6.0		50	25	12
NO_2	---	---	---		0	0	0
NO_3	---	---	---		0	0	0
Phosphorus (Total as P)	11 – 160	50	59		20	10	6
Grease (fat)	35 – 500	209	---		150	100	50
pH	*5.3 – 9.4	7.1	---		---	---	---

Note: *Industry values (6)

Source: EPA Technology Transfer Seminar, March 1973

Fats, Oils, and Greases (FOG)

In municipal treatment systems treating domestic and industrial wastes,

ordinances are usually adopted to protect treatment works by restricting
the concentration of FOG to less than 100 mg./l. Discharges to the
collection system in excess of 100 mg./l. would be prohibited or require
pretreatment. This particular requirement has been adopted widely for
municipal treatment systems largely due to the available "Model Ordinances" which serve as a guideline for drafting ordinances applicable to
specific municipal systems.

The principal difficulty experienced with this criteria is that the analytical
methods employed do not account for the wide variety of substances included
in the determination, i.e., any material which is hexane soluble and would
be subsequently evaporated with the hexane at 100°C., nor do they
distinguish between that matter which is of mineral origin versus the fatty
matter which may be of animal or vegetable origin.

What appears to further complicate the collective nature of the analytical
method is that there is no differentiation of the organic matter as to its
physical state, i.e., whether these materials are present in wastewater as
liquids or solids which may be readily separable by flotation, or whether
they may be present in finely divided states, emulsified or soluble and not
be readily separable. Also, no distinction is made as to the ability of the
wastewater treatment facilities to remove these substances with the usual
type of treatment afforded.

For example, although most municipal treatment plants have devices in
primary and now, secondary settling units to remove floatable substances
by retention baffles extended below the water surface with scum movement
and withdrawal provisions, very limited attention has been drawn to the
fact that fats, oils, and greases of animal and vegetable origin are treatable
biologically in both aerobic and anaerobic treatment units whereas FOG
of mineral origin are considered to be nonbiodegradable.

In order for biological degradation to occur, however, the FOG of animal
and vegetable origin must be in physical states which will permit the biological mass to be in contact with the material to be oxidized. Thus, if
the material separates, or floats on the surface of treatment units, the
opportunity for biological degradation is greatly reduced and this phenom-
enon has long attributed to the desired exclusion of these substances from
municipal treatment systems.

It is well known that anaerobic treatment systems such as anaerobic digesters
are particularly adaptable to the degradation of FOG from animal and vegetable origin, and are capable of higher degrees of volatile solids reduction
and greater volumes of methane gas production per pound of volatile solids
reduced than for other organic matter common to municipal wastewater
systems. This is not so for FOG of mineral origin wherein these substances

effectively coat insoluble surfaces and in high concentrations will effectively impair biological treatment by interfering with the normal mass
transfer functions of the biological system.

In instances where gross oil or greases are present in wastewater regardless
of the origin whether mineral or from animal and vegetable sources, these
substances which are readily separated by traps or limited gravity separation units, should not be discharged to a municipal collection system where
adverse effects of sewer clogging, excess accumulations in wet walls, or
overloading of scum removal equipment occurs.

For the discharge of dairy wastes to a municipal system, in that the FOG
are biodegradable, the principal concern should be directed to whether or
not the greases present will readily separate, cause sewer clogging, or
excessive accumulations. Floatable greases should be removed if these
adverse affects are noted; however, FOG in highly dispersed states, although in concentrations that may exceed the presently accepted level of
100 mg./l. should not be excluded from municipal treatment systems.

For example, it would not be very practical to require biological pretreatment of dairy wastes for the purposes of removing FOG of a dispersed nature
if the municipal system is going to utilize similar biological treatment processes. It is expected that there will be a concentration limitation imposed
on FOG of mineral origin and FOG of all types which are readily floatable.
Likely no restrictions would be placed on FOG in dispersed states of animal
and vegetable origin.

Municipal System Discharge

As indicated by others, the practice of discharging dairy wastewaters to
municipal systems is commonplace. In that dairy wastes are highly amenable
to treatment generally, the main concerns have been directed to the inter-
mittant nature of the waste discharges wherein the treatment plant may be
heavily loaded or subject to large variations for certain wastewater characteristics which may adversely affect treatment.

Because of the emphasis on establishing rate structures which reflect the
costs to the users of the system for appropriate Federal construction grant
monies, the dairy industry as with other wet industries is becoming more
waste conscious, employing in-plant waste saving devices and overall in-
plant improvements to minimize waste discharges. The characteristics of
dairy waste discharges have received some attention. The degree of attention is somewhat related to the percentage of dairy waste to total municipal
waste sources, wherein a high proportion of dairy wastes to total municipal
wastes appears to highlight the following waste characteristics:

1. The highly variable nature of dairy wastewater strength in terms of BOD may necessitate the employment of pretreatment in the form of aerated holding facilities. The ability of a municipal wastewater system to treat highly variable wastewater strengths depends upon the dilution and attenuation of the characteristic dairy wastes with wastes from other sources.

Of more concern are the resident times or detention times in the treatment units. A wastewater treatment plant that employs extended aeration having detention periods approaching 24 hours may realize little to no benefit in employing equilization or holding facilities prior to discharge to the municipal system. In other instances, where treatment detention periods are shorter, recirculation may assist but the overall effects must be observed and analyzed on a case by case basis.

2. Another dairy wastewater characteristic which draws considerable attention is related to the pH variation of the wastewater, particularly when peak alkaline conditions occur during cleanup operations. In that an upper limit of pH 9.5 has been adopted in certain sewer ordinances for the discharge of wastes to a municipal system, this value can be easily exceeded during periods of washing. Equalizing the waste discharge to attenuate the pH variation may be recommended but not always be warranted.

Whereas pH values below 5.5 may be harmful to sewers because of the corrosive nature of the wastewater, generally sewer construction of alkaline earth materials would not be so affected by high pH discharges. In terms of biological waste treatment, large variations in pH over a short period of time in the biological treatment unit would be undesirable. The variation of pH at the source of industry discharge may have little bearing on the pH in the biological treatment unit and, therefore, should be monitored in the treatment unit, not the discharge to the unit, to determine the pH range encountered.

Again, in activated sludge treatment systems with long detention periods, the pH variations are usually slight although large variations in pH may be evident in the incoming waste stream. Regulation of discharges to a municipal waste system and required pretreatment must be consistent with the desired end result sought.

3. With the increased emphasis on removing nutrients such as nitrogen and phosphorus from wastewaters that contribute to eutrophication or fertilization of lakes and streams, this may place an additional burden on contributors to the system particularly from point sources where the use of phosphorus bearing cleaning compounds are employed. Costs for treatment in some instances have been prorated on a per pound basis of P present in the waste discharges to the system. The methods employed for P removal usually result in high operating costs due to chemical additions required for P removal and

the handling of the resulting sludge with fairly nominal annual costs associated with capital improvements related to this removal function.

SPRINGFIELD, MISSOURI CASE HISTORY

A discussion of a joint treatment approach between a city (Springfield, Missouri) and several dairy products plants has been presented by P.T. Hickman of Hood-Rich Consulting Engineers of Springfield, Missouri. Springfield has grown phenomenally in both population and industrial activity in the past twenty years, and although there is a wide diversification of manufacturing, the largest single type of wet industry in the city is that of milk and milk product processing.

The city of Springfield and the surrounding area, being faced with the necessity of improving their wastewater facilities and preserving their unpolluted recreational water sites in 1955, decided to undertake a wastewater treatment system which would include the treatment of plant effluent from the city's manufacturing processes, rather than to encourage individual industrial treatment facilities with their inevitable multiple effluent discharge points.

The city is currently served by two wastewater treatment plants which utilize conventional primary treatment followed by the Kraus Modified Activated Sludge Process for secondary treatment. Both plants also employ separate anaerobic sludge digestion. The larger of the two plants, the Southwest Wastewater Treatment Plant, is designed to handle peak flows of 16 MGD with a BOD population equivalent of 310,000 (Figure 71). This plant accepts all of the industrial waste from the city with the exception of one milk bottling plant, which is tributary to the much smaller Northwest Plant.

Presently, the Southwest Plant is nearing both hydraulic and BOD capacity. Plans are now being developed to double the hydraulic capacity and to conform with recently adopted State Effluent Guidelines which call for a maximum effluent BOD of 20 mg./l. and NH_3-N (ammonia nitrogen) of 2 mg./l., along with disinfection, turbidity, and taste and odor requirements.

BOD removal will be done utilizing the pure oxygen process in an altogether new facility. The existing aeration tanks and air supply system will be used as nitrification tanks to convert ammonia to nitrates to meet the recommended maximum of 2 mg./l. NH_3-N. Nitrification will be followed by multimedia filters to remove excess suspended solids. Disinfection is to be done by utilizing ozonation which will also aid in turbidity and taste and odor removal. The modified plant design is shown in Figure 72.

FIGURE 71: CITY OF SPRINGFIELD, MISSOURI WASTEWATER TREAT-
MENT PROCESS (KRAUS)

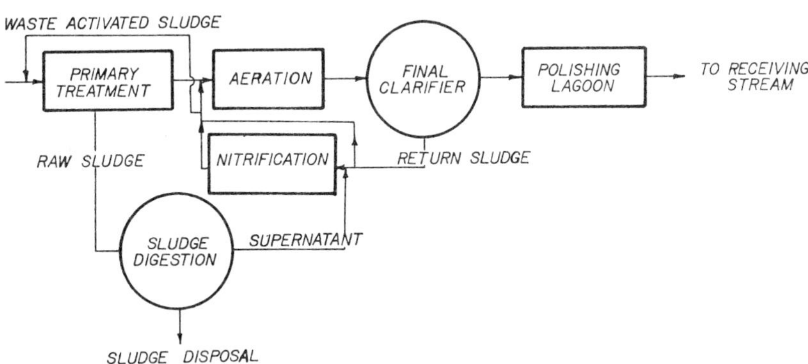

FIGURE 72: CITY OF SPRINGFIELD, MISSOURI WASTEWATER TREAT-
MENT PROCESS CURRENTLY IN DESIGN FOR S.W. PLANT

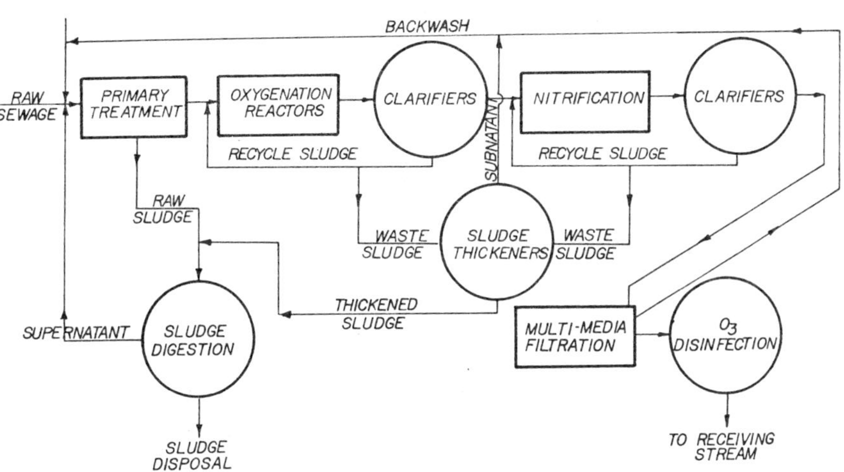

Source: EPA Technology Transfer Seminar, March 1973

In 1955, the citizens of Springfield voted nearly $10 million in bond funds
for sewage treatment improvement. One of the provisions was the establishment of a system of sewer service charges which included the provision for
surcharges to be levied against those industries that discharged wastewater

with BOD and/or suspended solids (SS) contents greater than domestic strengths. In May, 1971, Springfield citizens again voted bonds for major improvements to the sewer system. To finance these improvements an increase in the basic volume charge was made and the surcharge was increased to reflect the actual treatment cost experience averaged over the preceding five years for BOD and SS removal.

Although there are approximately 75 industries in Springfield which are classified under Section D of the Manufacturing Standard Industrial Classification, only 19 are subject to the industrial waste surcharge. The numbers and types in this category are:

Meat processing	6
Milk and milk product processing	5
Other food and kindred products	4
Commercial laundries	3
Pharmaceutical	1

Of the total dollar volume, the milk industry pays more than 55% of the surcharge receipts. The total load that these 19 industries place on the system and the treatment plants is: flow, 8%; BOD, 48%; and suspended solids, 18%. If one takes each one of these equally, the average is 25%. Also, if one looks at the total system income from sewer service charges, one observes that these 19 industries pay 25% of the total.

In a joint municipal-industrial system, the possibilities of having a waste that is amenable to biological, physical, and chemical treatment is quite good in comparison to individual industrial wastes. In Springfield's case practically all of the industry is of the food processing category whose wastes are readily treatable.

The joint treatment approach to wastewater treatment by the municipality and its industries has worked very well in the city of Springfield by producing sufficient revenue to finance the treatment on as nearly equitable basis as possible, and it is felt that the treatment is better than would have been achieved if the approach had been piecemeal.

WASTE TREATMENT ECONOMICS

The cost of disposal of dairy food plant waste is influenced by whether or
not the dairy food plant has its own treatment facilities, is utilizing an existing municipal facility and the ordinances and policies of the municipality.

As dairy plants become larger and as the capacity of municipal treatment
facilities become limited or marginal and as more stringent requirements
come into being in respect to effluents from treatment facilities, there
has been an increasing tendency to place an economic responsibility more
directly on the dairy food plants. A majority, approximately 90%, of the
dairy plants in this country utilize municipal treatment facilities.

As cities become more sophisticated and recognize the significance of the
dairy food plants to their BOD and COD load, there has been a trend to
either (a) charge the dairy food plant a surcharge on BOD, COD, suspended solids or other compositional factors that are above the concentration normally present in domestic waste or (b) arbitrarily limit the composition of the waste that can be discharged to the city system.

Several communities in this country, to meet imposed stringent stream
standards for BOD or to permit proper plant operation, have passed ordinances prohibiting the discharge of whey, placed a limit on BOD and/or
other components such as free lipids (hexane solubles) and/or suspended
solids. These limits have been based on domestic waste values and have
generally been formulated in terms of a ppm strength rather than an absolute quantity. Generally the limits set to date have ignored the quantity
of wastewater and focused only on the ppm strength. It would appear to
be more logical to take into consideration both wastewater volume and
per unit volume strength and to establish standards on the basis of coeffi-
ciency strength, such as the total pounds of BOD per day or the total

pounds of free lipids per day. Where limits have been set for the BOD at
200 to 250 ppm, dairy plants have been forced to either move their operation or to install their own pretreatment facility. With current technology, eliminating all possible sources of BOD through various control procedures, it would appear that 400 to 500 ppm is the lowest achievable
unit volume strength that can currently be obtained. As more and more
cities impose such standards, there will be an increasing demand to install
pretreatment facilities in dairy plants. The need for such standards on
effluents to municipal systems needs very careful study.

During the course of an EPA study (2) there was an occasion to work with
one dairy plant facing a limit in its BOD and suspended solids of 200 ppm
to be achieved within 18 months. This plant, manufacturing milk and cottage cheese, produced about 100 pounds of BOD per day at a concentration
of 2,000 ppm, if whey was excluded from the wastewater. Using all the
methods outlined earlier in the report for conservation resulted in a reduction to 550 ppm. They considered using dilution to bring the BOD down
to 200 ppm limit but finally decided to install a treatment facility.

This same problem will be facing dairy food plants increasingly in the next
five to ten years. As more demands for the development of waste treatment
facilities are made by dairy plants, there will be a need to develop more
basic information and efficient treatment systems for fluid plant waste.

The recent actions taken by Government such as the rules and regulations
established by the EPA-OWQ regarding grants for construction of treatment
works make for a greater burden than previously shared by industrial users
of new municipal treatment works (63). The rules include the following:

1.   The applicant shall assure the Commissioner that each ap-
     plicant will require pretreatment of any industrial waste
     which would otherwise be detrimental to the treatment
     works or its proper and efficient operation and maintenance.

2.   An equitable system of cost recovery must be implemented.

3.   The cost recovery system shall produce revenues in pro-
     portion to the percentage of industrial wastes, relative to
     the total waste load to be treated including operation,
     maintenance, amortization and other additional costs.
     For the dairy industry this means larger sewage bills and
     possible elimination of the practice of dumping whey into
     the municipal sewers. All of the above demonstrates the
     need and urgency of control and management of water and
     wastewater in dairy processing.

New State and Federal regulations are requiring higher standards of treatment of municipal and industrial wastes as noted above. Rigid enforcement of the new and prior legislation has forced all treaters of industrial wastewater to more advanced methods of treatment. State and Federal governments help local governments in financing new treatment facilities or additions to existing treatment systems. However, Federal grants, along with State matching grants, require that industries be charged "equitable" amounts for water and sewer service.

The problems associated with implementing the new regulations and funding the higher costs of these environmental improvements is left to local governments or individuals. A common method for municipalities is to first pass an industrial user ordinance. A method usually presented in such an ordinance for meeting Federal requirements for "equitable" charges is a system of industrial waste surcharges. Surcharges are based on the pounds of waste discharged and/or on the volume of effluent discharged. A surcharge is levied in addition to the normal sewer charges based on water used.

Sewer charges are normally a constant proportion of an industrial firm's water bill and examples noted include 10 to 200% of the water bill with a normal range of 50 to 100% of the water bill. The general theory of surcharges is that they allow a town to recover some revenue while providing industrial firms an economic incentive to reduce the amount of water carried waste.

Industrial surcharges are expected to (64):

1.   Provide a more equitable recovery of the costs of waste treatment than ordinary charges based on volume alone.

2.   Reduce wastes discharged to the city sewers because industrial firms are required to expend capital to combat excessive surcharges.

3.   Reduce water used by industry because water reductions usually precede waste reductions.

4.   Allow a treatment system to operate at this lower waste loading with the same effect as expansion of facilities, requiring new capital investment.

SURCHARGE

What Is a Surcharge?

A charge based on the pounds of waste material in industrial wastewater in excess of normal levels of concentration is called a "surcharge." It is

levied in addition to the normal sewer service charge which is the regular
charge for treating normal strength wastes and is based on volume alone.
A surcharge is normally only for pounds of waste above normal and an economic incentive is provided to reduce the strength of the wastes. Some
North Carolina cities with surcharge ordinances include Charlotte, Greens-
boro, Winston-Salem, Monroe and Durham. A surcharge ordinance passed
in Raleigh went into effect on May 1, 1973.

What Are Normal Wastes?

Normal waste refers to that waste normally found in wastewaters from
households. These waters usually have the following composition:

BOD	250 to 300 mg./l.
COD	250 to 370 mg./l.
SS	50 to 200 mg./l.
Hexane Solubles	100 mg./l.

Harper et al (2) found the waste composition for 10 selected dairy plants
processing at least 250,000 pounds per day of milk to be:

	Range	Average
BOD	750 to 4,200	3,100
SS	200 to 3,700	2,450

Since dairy wastes are more concentrated than normal wastes, surcharges
are being included in billings to the dairy industry.

Measurement of Flow and Strength

The main weakness in many surcharge ordinances is in the method of
sampling. Someone must determine the waste concentration of a plant's
effluent and also determine the flow. Flow is often determined from water
readings which is disadvantageous to the normal dairy plant. On an average, the effluent in a study (14) was 73.67% of the water received from the
city. At least one municipal ordinance in the state will allow a plant credit
for their consumptive uses in product, cooling tower evaporation and steam
losses. The well managed milk, cottage cheese, and frozen products plant
should have a consumptive use of at least 20 to 25% of their water received
on a yearly average.

A representative sample of dairy plant waste is almost impossible to obtain.
The wide ranges presented in a study (14) show that day-to-day variations
are too large to allow a sample from one day to be representative. Seasonal
production of certain items surely eliminates one set of samples being

representative of a dairy processing plant. Also, the only true measure of
the waste from a plant is a 24 hour proportional, composite sample. This
is a sample that mixes volumes of waste in proportion to effluent flow over
a representative time period. Night flows from dairy plants are largely
cooling waters and leaks and are not heavily polluted. A sample taken
from 8:30 to 4:30 is not representative of the whole processing day.
Sampling by proportional, composite sampler requires a flow regulating
device such as a weir or flume and a sampler such as the arm scoop type.
Such techniques are accurate but are expensive. Most muncipalities and
industries are not willing to go to such extremes to get a representative
sample.

Chemical analyses must be run on the sample of wastewater to determine
the waste characteristics. These tests should be performed on fresh wastewater samples which have been refrigerated during collection. Failure to
follow standard analysis procedures can result in erroneous test results. Well
trained, competent technicians are especially important.

Cost of Surcharge

Figures 73 and 74 illustrate potential water and waste charges for a dairy
plant examined in this study (14). Figure 73 presents the surcharge costs
a dairy plant will incur just from the processing of cottage cheese. Note
that the total surcharge cost is approximately $4,000/month for a produc-
tion using 20,000 gallons of milk each day. Water and sewer costs are another $537.72/month with a total cost per pound of cottage cheese produced
of 0.9¢/lb.

These figures are for a plant processing 20 days each month. This cost is
a major item in the cost of production of the cottage cheese. However, if
the surcharge rate goes up from $0.0354/lb. BOD and TSS to $0.0739/lb.
BOD and TSS, the cost per pound of cottage cheese produced goes up to
1.7¢/lb. Dairy processors who produce cottage cheese must realize the
implication of surcharge costs on the profitability of their finished product.

Figure 74 presents the costs for a combination plant as presented in the
study of the Pine State Creamery (14). The waste generated and water use
averages for this study were used to prepare the graph. Therefore, the
curves are valid for this particular plant and are not necessarily representative of any other plant. A plant processing an average of 500,000
pounds of product each day for 22 days of the month will have a monthly
municipal bill of approximately $10,000. This represents a cost to the
plant of 0.09¢/lb. of product or approximately 0.8¢/gal. of product.

FIGURE 73:   SURCHARGE FOR DIFFERENT LEVELS OF COTTAGE CHEESE PRODUCTION

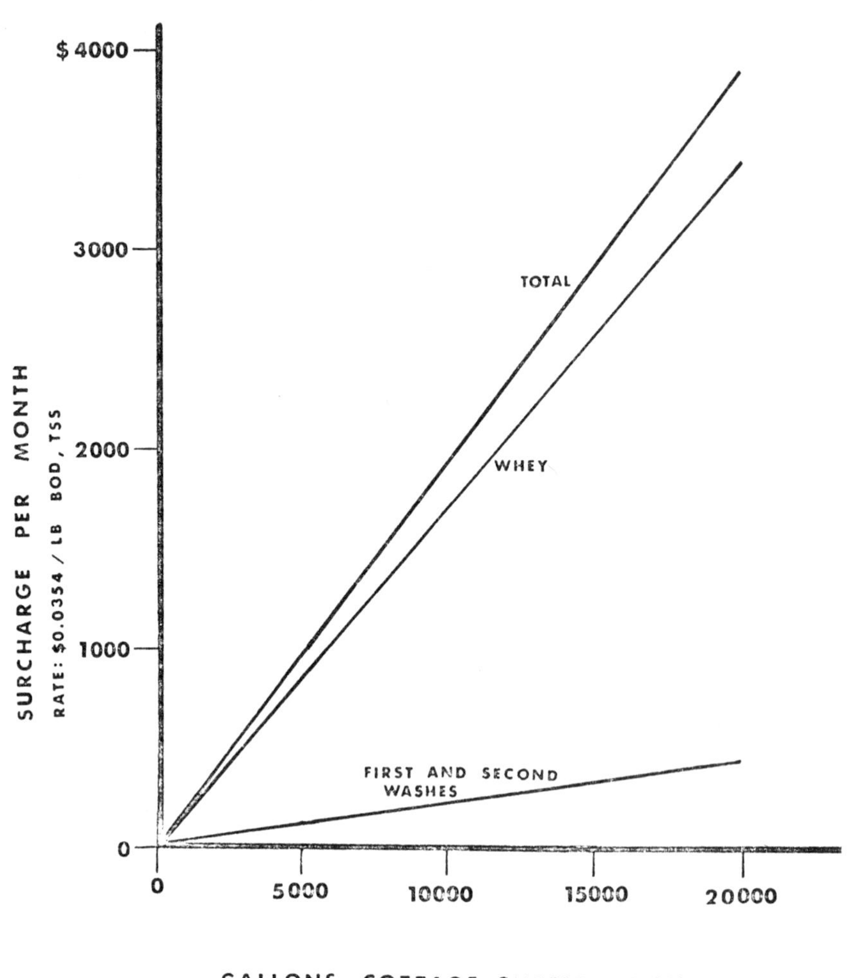

Source: Report PB 220,704

FIGURE 74:   PREDICTED MUNICIPAL CHARGES BASED UPON WATER USE AND WASTE GENERATION AT PINE STATE CREAMERY

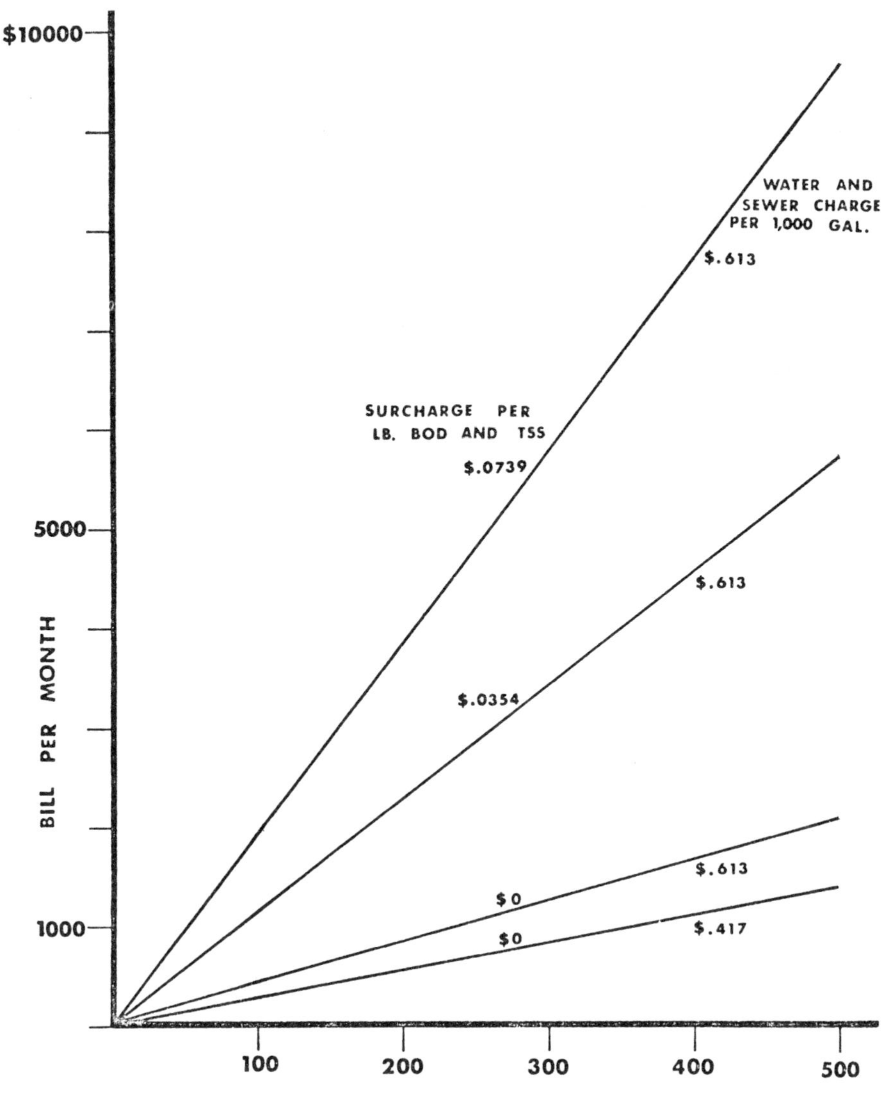

Source: Report PB 220,704

Effect of Surcharge

Most dairy plants will be forced to reduce their wastes to avoid much of the surcharge costs. Segregation of concentrated wastewater such as whey and subsequent disposal by swine feeding or surface application will be required. Representative formulas for surcharge calculation are as follows:

1. Surcharge = 0.9 x BOD in ppm ÷ 250 x 8.34 x gallons of wastewater x the rate for water usage.

2. Surcharge = ($0.023 per 1,000 gallons x 0.014 x lb./BOD_5) + (0.023 per 1,000 gallons x 0.024 x lb. susceptible solids) – ($0.004 per employee day work) – (10,000 gallons x 0.023 + lb./BOD in 10,000 gallons x 0.014) + (lb. suspended solids in 10,000 gallons x 0.024).

3. Surcharge = B + C – D + E where B is the cost factor for total suspended solids above established minimum of 250 ppm. C is the cost factor for oxygen demands in excess of 250 ppm BOD or COD; D = cost deduction factor for that portion of water that goes into the manufactured product or its production by-products or is disposed of by other means than the city sewers; E is the added cost factor for properly metered private water ultimately discharged to sewers, with the cost being determined of $0.02 per lb. for BOD_5 above the 250 ppm + 2.0 x lbs. of suspended solids above the 250 ppm limit.

4. Surcharge = (suspended solids in ppm – 350 x 8.34 x F x 182.5 x 0.95 x 40)/2,000 + (ppm BOD – 300 x 8.34 x F x 182.5 x 0.95 x 45)/2,000 where 350 is the total suspended solids in normal sewage, 300 is the BOD in normal sewage, 8.34 is the weight per pound in a gallon of water. F is the million gallons of wastewater per day; 182.5 is the number of days for each semiannual period; 0.95 is the factor allowance for 95% purification; 40 is the surcharge in dollars per ton of excess suspended solids ($0.02/lb.) and 45 is the surcharge in dollars per ton of excess five-day BOD ($0.025/lb.)

5. Surcharge = $0.0129 x (ppm BOD_5 – 300) x gallons x 8.34 + $0.21 per 1,000 cubic feet of water consumed.

6. Surcharge = a fixed monthly rate x (parts per million BOD – 250) x 0.0025.

7. Surcharge = 1.5 x dollar value of water bill + $0.01/lb.
BOD and $0.01/lb. suspended solids in excess of 250 ppm.

On the basis of BOD surcharges, the communities from which information
was obtained vary in their rate of surcharge for BOD and suspended solids
from $0.01 to $0.08 per pound. In the survey conducted during an EPA
study (2), cost of wastewater treatment was available for 20 of the commercial dairy plants.

TABLE 70: WASTE TREATMENT COSTS FOR COMMERCIAL DAIRY FOOD PLANTS IN 1970

Plant No.	Pounds BOD per Month	Method of Treatment	Cost/Month (1970)	Explanation of Charges
3	20,900	City	$ 958	$5,000 annual cost + $5.12/1000 gpd > 50,000 pd + $1.70/lb BOD > 200 ppm
4	58,600	City	700	1.5 x amount of water bill + $0.01/lb BOD and $0.01/lb suspended solids in excess of 250 ppm
5		City	100	Unknown
6	35,000	City	444	Volume x water rate
7	34,540	Plant Act. Sludge	150	Operating expenses
8	83,600	Plant Act. Sludge & City	5,500	Operating cost and amortization of treatment plant + charge-for city treatment
10	21,000	City	262	Volume waste water x 60% water costs
11	4,000	City	6,300	Based on water usage x $53.27/mg % BOD
14	-	City	304	$0.047/1000 ft^3 of water purchased
16	-	City	1,824	$0.23/gallon
19	-	City	48	Based on water usage
	36,850	City	600	Based on water usage
23	34,364	City	5,413	Based on water usage x $53.27/mg % BOD
24	2,574	City	70	Based on water usage
26	-	City	165	Based on water usage
27	1,770	City	$ 250	77% water bill (volume x 8.34 x (0.0178/ppm BOD - 250) + (0.0170 (ppm SS-300)
28	19,690	City	1,146	77% water bill (volume x 8.34 (0.0178 (ppm BOD - 250) + (0.0170 (ppm SS-300)
30	134,860	City	5,416	Water rate + 0.025 (ppm BOD-250) + (ppm SS-300) x gal x 8.34
31	62,200	City	2,916	90% water bill x 0.02 (ppm BOD-250) + (ppm SS-300)
51	64,800	City	3,000	Water rate + 0.02 pound BOD over 200 ppm + 0.02 pound SS over 200 ppm
57	116,760	City	12,500	Fixed charge (ppm BOD-250) (0.0025) on water usage

Source: EPA Report 12,060 EGU, March 1971

The information is tabulated in Table 70 in respect to the pounds of BOD produced per month calculated on the basis of days of operation per month times pounds of BOD per day, the cost per month in explanation of the charges from which the cost is based. For those plants paying a surcharge the cost ranged from $250 per month to $12,500 per month, which represents costs of $3,000 and $150,000 per year. The plant paying wastewater treatment costs of $150,000 per year is paying approximately $0.05 per pound of BOD and discharging over 100,000 pounds of BOD per month; whereas the plant paying $3,000 per year is discharging less than 2,000 pounds of BOD per month and paying about $0.01 per pound.

In discussing the difference in the rate per pound of BOD with sanitary districts, they attributed the variation in cost across the country to differences in capital investment, operating costs and the relative load on the plant from industrial sources. In the case of the plant paying $3,000 per year for wastewater treatment, it contributes less than 5% of the BOD load on municipal treatment system; whereas the plant paying the largest fee is contributing approximately 80% of the BOD load to the municipal treatment plant.

Costs for plants paying a "cost only on water usage" ranged from $48 per month to $6,300 per month reflecting a marked difference in the cost of water in different portions of the country. Most of the plants operating their own treatment facilities did not have accurate cost data. In one case where the plant was treating about 80,000 pounds of BOD per month, the cost was $5,500 per month, which included a portion of the capital investment. The discharge from the treatment plant was being treated by the city plant and a charge was being assessed over a given quantity of BOD.

COST OF WASTEWATER TREATMENT FACILITIES

Using the value of $0.02/lb. BOD as a reasonable mean figure for surcharge, the surcharge cost, which represents only a portion of the overall charge for municipal waste treatments attributal to plants treating from 100,000 to 700,000 pounds of BOD per day, is shown in Figure 75. This figure shows the cost of a normal receiving, filling, pasteurizing and over-all fluid milk plant with a 1.45 pound BOD coefficient per 1,000 lbs. of milk processed and the cost for a plant operating at a coefficient of 3 lbs. BOD per 1,000 lbs. milk processed. Figure 76 is a continuation indicating the cost for 5 lbs. BOD per 1,000 lbs. cottage cheese wash water containing 8 lbs. BOD per 1,000 lbs. milk processed, and cottage cheese whey containing 26 lbs. of BOD per 1,000 lbs. milk processed. In many instances the surcharges shown represent only about 50% of the over-all waste treatment charge. In addition to reduction of the BOD coefficient it is equally significant to reduce the water volume.

FIGURE 75: COST OF BOD SURCHARGE FOR DIFFERENT STANDARD UNIT OPERATIONS AND FLUID MILK WITH DIFFERENT LEVELS OF BOD AS A FUNCTION OF MILK PROCESSED

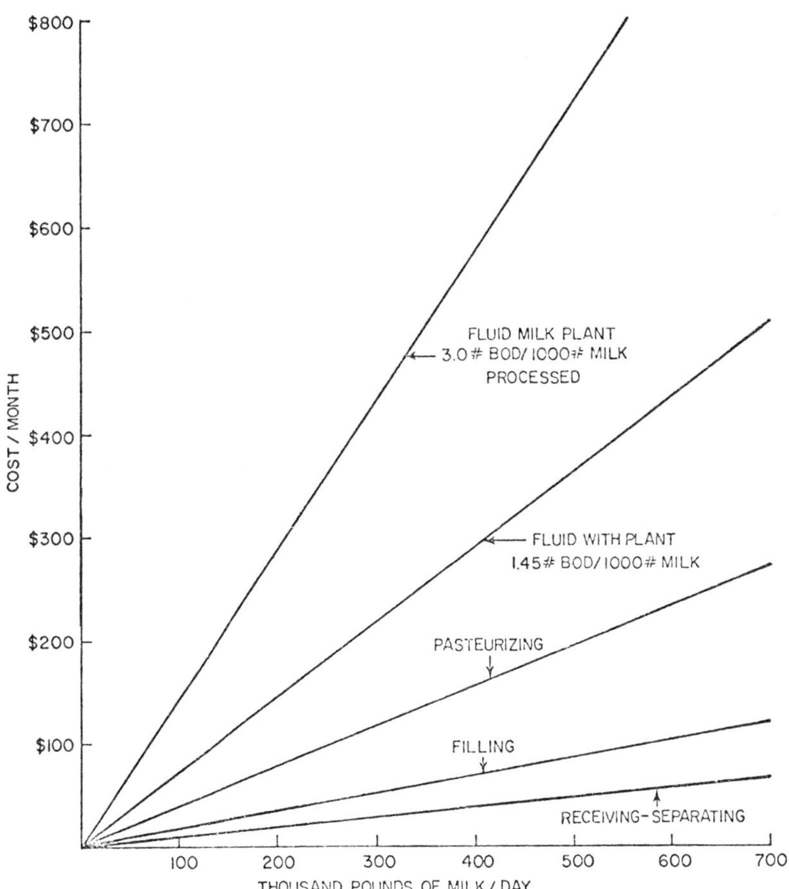

Source: EPA Report 12,060 EGU, March 1971

FIGURE 76:   COST OF BOD SURCHARGE FOR WHEY, COTTAGE CHEESE WASH WATER AND FLUID MILK PLANTS WITH DIFFERENT LEVELS OF BOD AS A FUNCTION OF MILK PROCESSED

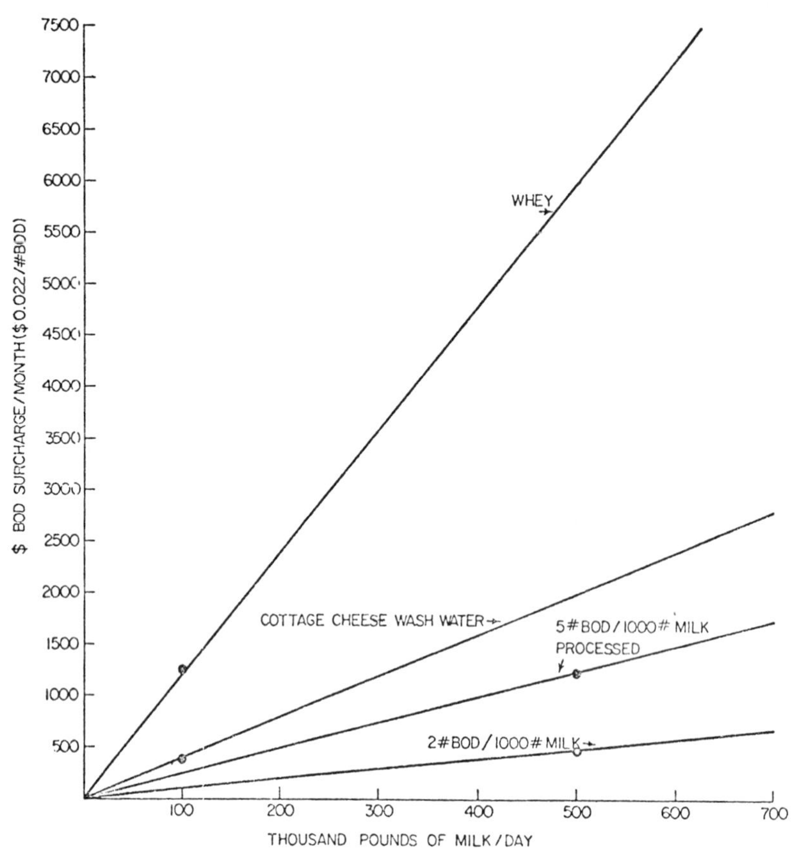

Source: EPA Report 12,060 EGU, March 1971

The attainment of actual cost information on the construction and operation of dairy waste treatment facilities is difficult, based in part upon the
lack of information in industry and at the same time the contradiction of
information available in the literature. Zall (65) in his investigation,
found that only one of 55 food plants visited in 1967 knew waste discharge
volumes and only 16 had any idea of cost of waste disposal. In the Ohio
State survey (2) of 57 plants, only 20 had actual cost figures.

In respect to operations the cost stated in Zall's study for the lowest individual plant per year was $1,500 and the highest cost was $25,000. The
operating costs were figured on the cost of man-hours expended by part-
time labor assigned in the sewage disposal area, the cost of chemicals used
in the treatment system, dollar purchases of material to promote digestion
and power consumption assigned to electric motors in the waste treatment
plant. In Zall's investigation, and also in the Ohio State survey information, neither physical size of the food plant nor its annual dollar sales cor-
related to waste handling expenses. The operating costs for different methods of treatment follows:

Method of Treatment	Dollar Expenditure per Year (1967)
Spray irrigation	$ 1,500
Spray irrigation, excluding whey discharge	12,500
Spray irrigation, including whey discharge	15,000
Oxidation pond	1,500
Oxidation pond	10,000
Aerated lagoon	8,500
Activated sludge	10,000
Municipal activated sludge with BOD surcharge	6,000
Municipal activated sludge with BOD surcharge	25,000

The annual operating costs differed widely as noted in the Ohio State survey data (2).

The operating costs found in the Waste Profile 9 study (3) are indicated in
Table 71. These are presented for a typical technology large plant in dollars (1966) per thousand gallons of wastewater treated. The ridge and furrow and spray costs were essentially the same from plant type to plant type
and unrelated to the BOD load. Aerated lagoon costs ranged from $5.12
to $252.88. These costs varied widely between different plants with an
apparent general relationship between the pound BOD per 1,000 pound
wastewater and the cost of operation of the aerated lagoon, which would

reflect the cost of oxygen incorporation which should be larger as the BOD loading per unit volume increases. In respect to trickling filter operations the costs range from $121 to $1,024, and appeared to be unrelated to the type of operation or BOD load per unit volume. The same observation is generally true in respect to the activated sludge costs which ranged from $80 to $682 a month.

TABLE 71:   OPERATING COSTS FOR THE INSTALLATION OF VARIOUS TREATMENT METHODS FOR DAIRY FOOD PLANT COSTS (1966 DOLLARS)

Method	$/1,000 Gallons of Wastewater Treated				
	Butter	Cheese	Condensed	Ice Cream	Milk
Ridge and furrow	59.50	73.17	68.20	67.94	54.56
Spray irrigation	163.35	219.50	192.66	192.30	133.20
Aerated lagoon	98.48	243.90	18.16	5.12	252.88
Trickling filter	170.60	1,024.39	121.10	291.02	194.12
Activated sludge	44.61	682.92	80.73	192.44	128.31
Pound BOD/1,000 gallon wastewater	9.32	16.1	1.45	0.39	25.5

Source: FWPCA Publication IWP-9

The information on capital costs is somewhat confusing also. An accurate assessment or development of the universally applicable formula for determining the capital operating costs of dairy food plant waste treatment facilities are difficult because of contradictory information, limited nature of the data available and major differences in the relative costs involved in treatment facilities in different parts of the country. Labor costs, construction costs and land costs are major contributing factors to variability. As noted, the most complete presentation of cost figures, those in Waste Profile 9 study (3) appear to be higher in their estimations of cost than almost all other reports and appear to be based to a large degree on the early work by the U.S. Department of Agriculture and the University of Pennsylvania.

Values, when corrected to current costs, appear to be in general agreement around the world and are in agreement with information obtained from consulting organizations who have had experience with the construction of dairy food plant waste treatment facilities. Based on information obtained from consulting companies in the North-Central state area, the following formulas are suggested as approximate estimates of cost taking into account both hydraulic and organic loading.

Construction Costs

Double aerated lagoon:
$$\text{Cost} = [\$200 \times \text{waste volume coefficient} + \$175 \times \text{BOD coefficient}] \times \frac{\text{gallons wastewater}}{1,000}$$

A single recirculating trickling filter:
$$\text{Cost} = [\$250 \times \text{waste volume coefficient} + \$200 \times \text{BOD coefficient}] \times \frac{\text{gallons wastewater}}{1,000}$$

A modified activated sludge system with 24 to 48 hour retention:
$$\text{Cost} = [\$300 \times \text{waste volume coefficient} + \$275 \times \text{BOD coefficient}] \times \frac{\text{gallons wastewater}}{1,000}$$

Operating Costs

The current operating costs, exclusive of labor, would approximate 75¢/1,000 gallons and 50¢/pounds of BOD for aerated lagoons; $2.50/1,000 gallons and $3.00/1,000 pounds of BOD for activated sludge; and trickling filter operating at $2.00/1,000 gallons and $2.50/1,000 pounds of BOD.

The information in respect to the cost of constructing and operating waste treatment facilities for dairy food plants is based on limited and insufficient information for full utility to the industry. At the present time each particular installation will have to be figured according to current local circumstances.

CASE HISTORIES OF
DAIRY WASTEWATER TREATMENT

STOCKTON, ILLINOIS

The Kraft Foods Division facility at Stockton, Illinois consists of two
separate operating plants, one for the manufacturing of bulk Swiss cheese
and the other for processing whey (13). All of the whey from the cheese
operation is condensed at the whey plant with the majority of it being
spray dried along with condensed whey received from other Kraft manufacturing facilities in the area. The balance of the whey is processed
into several other institutional and industrial products.

As is indicated in Table 72, the wastewater volume has increased at
Stockton from about 86,000 gpd in 1968 to 110,000 gpd at present. Over
this same period, however, the BOD load leaving the plant has been reduced from 1,950 to 900 pounds per day by in-plant process modifications
and improved housekeeping.

The existing waste treatment system has evolved over the almost 60 years
that the plant has been in operation through efforts to solve the wastes
problem in the simplest and most economical manner. Summer operation
consists of pumping the total plant discharge through the irrigation system
on site summarized in Table 73. The concept here is through the use of
automatic controls to rotate the dosing of the land through a preplanned
application cycle so overdosing does not occur at any one location and a
maximum of percolation, evaporation and transpiration of moisture will
occur.

Thirty-two nozzles are installed on site, each having a capacity of 90 gpm,
and dose in rotation. Each has the ability to spray roughly an acre. By
the use of this approach, maximum advantage is being taken of the land

and its cover crop to absorb the total plant discharge.

TABLE 72:   WASTE LOAD BEING DISCHARGED BY STOCKTON PLANT*

	WHEY PLANT		CHEESE PLANT		TOTAL	
	1968-9	1972	1968-9	1972	1968-9	1972
Volume gpd	57,000	80,000	29,000	29,000	86,000	110,000
BOD mg/1	2,900	900	2,400	800	2,700	870
BOD Load lb/Day	1,370	600	580	300	1,950	900

* Significant reduction in the BOD load due to in-plant processing

modifications and improved housekeeping.

TABLE 73:   THE KRAFT FOODS IRRIGATION SYSTEM AT STOCKTON

SPRAY FIELD

        50 acres of moderately sloping ground

        32 operating sprayheads + 2 standby

        Volume - 90 gpm (each)

        Coverage - 220 ft. diameter circle (approximately 1 acre)

MODE OF OPERATION

        Automatic

        16 circuits - 2 sprayheads/circuit

        Rotates through 16 positions

        Preset at 90 minutes per position

        Mowed - every 3-4 weeks with 6 foot chopper mower

Source: Watson, K.S., EPA Technology Transfer Program, March 1973

In the wintertime an activated sludge system is used. A summary of information on the activated sludge system is shown in Table 74. The basin is
operated using diffused air from a bottom diffuser normally supplemented
by two floating aerators. The discharge from the basin is passed through a
clarifier. From the clarifier, final disposal is made to a ridge and furrow

system on site. This system is on contour with an overflow of the control
gate of one trench to the adjoining trench below.

TABLE 74:   WINTER SYSTEM FOR HANDLING EFFLUENT AT STOCKTON

Activated Sludge Plant

  Aeration basin: 212' x 68' x 10'

    Aerators   - 3 - 40 HP blowers each delivering 400 scfm @ 4.5 psig

                2 floating aerators

    Volume   - 570,000 gallons

    Capacity   - about 1.3 lbs. O_2/lb BOD applied

  Clarifier: 24' diameter x 9' SWD - 1/2 HP sludge rake

    Volume   - 30,000 gallons

    Detention - 7.2 hrs.

  Sludge Lagoon: 100' x 40' x 10'

    Volume   - 300,000 gals.

    Supernatant fed back to aeration basin

Ridge and Furrow System

    10,000 lineal feet on contour

    18" deep by 4' wide furrow (trench) separated on 12' centers by ridge

    Overflow control gates at ends of adjoining trenches

Source: Watson, K.S., EPA Technology Transfer Program; March, 1973

The usable volume of the basin is 570,000 gallons having a detention time
of five days. It is loaded at about 24 pounds of BOD per thousand cubic
feet. When sludge must be removed from the system, it is drawn from the
clarifier to a sludge lagoon located nearby. Sludge is removed from the
lagoon once a year, in the spring, and spread on adjoining grassland.

For the plant to discharge into the small stream to which it is tributary and
meet the very stringent requirements of the Illinois EPA, the effluent concentration would have to be less than 5 mg./l. of BOD and 4 mg./l. of
suspended solids. The decision was therefore reached that necessary facilities should be installed to retain the effluent on site and depend upon
evaporation and percolation for disposal. For these reasons two impoundment

and flood control ponds were located at the lowest point on the property.
These lagoons of 3,100,000 and 4,250,000 gallons capacity were provided
to catch any runoff from the site during either summer or winter operations.

In relating how this plant solved its problem, the point should be made
that the solution has been tailored through evolution to this particular
plant's waste load in the particular habitat where located. Many processes
are in use here, however, which could have application to other dairy
plant wastes. The soundest and most economical solution to a plant problem will always be arrived at by tailoring the treatment to the particular
situation in question.

NORWICH, NEW YORK

Coverage of this plant is described (13) because its treatment plant is new
and of the biological type which is widely used in the handling of dairy
plant wastes. This Sheffield Chemical manufacturing facility is not a
typical dairy plant, but its principal raw material is whey concentrated
to 50% solids. Among the major products manufactured by this Norwich
facility are lactose, calcium lactate, sodium caseinate and food flavorings.

When it became apparent that the lagoon system being used at Norwich
would have to be replaced by more efficient treatment facilities, agreement was reached with the State of New York that some pilot scale work
should be done to provide the basis of design for the new treatment facilities. Consequently, a biological pilot scale unit was designed and placed
in operation in October, 1970. The pilot facility was, and can still be,
operated on a side stream of the discharge from the manufacturing plant.

The pilot facility was operated for about nine months under laboratory control. During this same period, the design of the full-scale treatment facility was moved forward as was a program of in-plant, short-of-treatment
steps to better manage water and reduce the waste load. Construction of
the waste treatment facility was started in July, 1971 and it was placed
in operation in February, 1972.

The full-scale facility consists of an activated sludge basin broken into
three compartments, in which the flexibility has been incorporated to
permit it to be operated using either the contact stabilization or extended
aeration mode. The biological treatment basin is followed by a final
clarification stage consisting of two clarifiers. Alum is fed to the aeration
basin effluent to improve the flocculation in the final clarifier. When
sludge must be drawn from the system, it is removed from the clarifier to
a sludge tank to be scavenged from the site. Pumps, blowers, the flow
recorder and the laboratory are located in a control building.

The treatment facility was designed to handle a hydraulic loading of 250,000 gallons per day and a BOD load of 2,500 pounds per day with 90% removal of BOD anticipated. A maximum retention time of 3.2 days was provided in the aeration basins. Additional information on sizes and specifications of equipment is shown in Table 75.

Since the level of suspended solids in the manufacturing plant effluent did not appear of great significance, no primary treatment was incorporated in the system. Since the need for feeding nutrients to improve the operation of the system had not been established prior to its being placed in operation, ammonia is being fed on an improvised basis into the influent end of the aeration basin. Since the use of ammonia is beneficial, facilities are being added to the system to permit ammonia to be fed on a permanent basis.

TABLE 75: SIZES AND SPECIFICATIONS OF WASTE TREATMENT FACILITIES AT SHEFFIELD CHEMICAL, NORWICH

Aeration Facility

1. 3 tanks each 160 ft. long, 16 ft. wide, 11 ft. liquid depth, 28,160 cu. ft. volume.
2. Loading - 30 lbs. BOD per 1,000 cu. ft. of aeration tank volume.

Air

1. 1.5 lbs. O_2 per lb. BOD design.
2. 3 blowers each 40 hp., 1,200 cfm.
3. 300 coarse bubble air diffusers.

Settling Tank

1. 17 ft. diameter, 3,600 cu. ft. volume.
2. Retention time 2.9 hours at 1.1 mgd.

Source: Watson, K.S., EPA Technology Transfer Program; March, 1973

The concept in use at this plant is to completely separate the process wastes from the cooling water so the former can be put through the treatment plant and the latter bypassed around same. Thus, sampling locations have been provided for ahead of, and following, the treatment facilities. Finally, a sampling and measuring station has been provided for monitoring the combined process and cooling wastewater stream which flows into the river.

Some feel for the operation of the treatment facilities can be obtained by reviewing Tables 76 and 77. Operating experience, to date, indicates that the 90% reduction in BOD for which the system was designed can generally be met. As is borne out by low minimum effluent concentrations shown in Table 77, at times the system does go above 90% efficiency in removal of BOD. Operating experience has further demonstrated that the extended aeration mode of operation produces considerably better results than does contact stabilization.

TABLE 76:   MEAN RESULTS OBTAINED FROM THE SHEFFIELD CHEMICAL TREATMENT FACILITY AT NORWICH*

Week Ending	Flow GPD	BOD mg/l Influent	Effluent	COD mg/l Influent	Effluent	pH Influent
10/6/72	171040	907	80	33281	369	7.4
10/13	222860	1200	54	3104	754	7.8
10/20	185780	1276	62	3047	241	7.1
10/27	153200	1983	52	4075	70	6.2
11/3	180933	1639	94	5425	312	7.4
11/10	146120	1570	18	1588	433	6.4
11/17	169920	1023	21	2533	61	6.2
11/24	154480	871	21	2945	74	6.5
12/1	138140	1895	9	5307	170	5.0

* Based on daily analyses (5-day week) of representative composite samples

TABLE 77:   BOD RESULTS OBTAINED FROM THE SHEFFIELD CHEMICAL TREATMENT FACILITY AT NORWICH*

Week Ending	Influent mg/l Max.	Min.	Effluent mg/l Max.	Min.
10/6/72	1608	257	142	16
10/13	2248	791	74	21
10/20	1570	778	101	15
10/27	3621	1248	75	29
11/3	2231	1253	189	16
11/10	2873	832	24	12
11/17	1279	832	40	4
11/24	1005	736	31	12
12/1	2196	1475	18	3

* Based on daily analyses (5-day week) representative composite samples

Source: Watson, K.S., EPA Technology Transfer Program; March, 1973

A program of relating river conditions to treatment plant operation has been launched and will be expanded. An initial look at river conditions, on the basis of dissolved oxygen, is shown in Table 78. It will be noted that as a general rule the dissolved oxygen in the stream is higher below the plant discharge than it is above. Just recently, the pilot facility was placed back in operation to develop a better feel for how to upgrade the performance of the full-scale unit.

TABLE 78:   RIVER CONDITIONS NEAR SHEFFIELD CHEMICAL TREAT-
MENT FACILITY AT NORWICH*

Week Ending	Above D.O. mg/1	Below D.O. mg/1
10/6/72	8.8	9.4
10/13	9.4	9.8
10/20	9.9	10.2
10/27	9.7	10.1
11/3	10.1	10.7
11/10	9.0	10.3
11/17	10.9	11.5
11/24	10.5	10.4
12/1	12.7	13.0

* Based on daily analyses (5-day week) of grade samples

Source: Watson, K.S., EPA Technology Transfer Program; March, 1973

SOUTH EDMESTON, NEW YORK

This plant of the Breakstone Sugar Creek Foods Division produces yogurt and ricotta cheese. The plant effluent is treated on site and discharged into the stream. The wastewater effluent from the processing plant is perhaps somewhat lower in concentration than is the case with that from many dairy plants. From Table 79 it is apparent that the average BOD concentration is 485 mg./l. and the suspended solids 147 mg./l. As is indicated by the analytical characteristics, every effort is made to keep whey out of the plant discharge.

The effluent treatment system consists of a raw wastes pumping station, aerated lagoon, clarifier, sludge basin and chlorinator. A flow diagram of the system is shown in Figure 77. In the center of the aeration basin is located a 20 hp. high speed floating aerator.

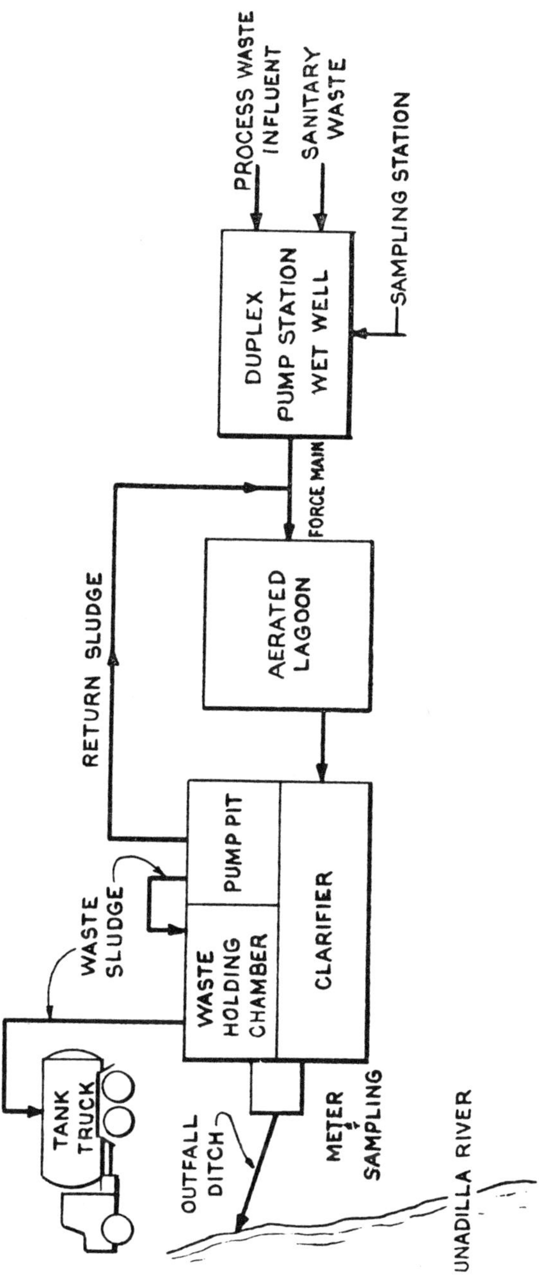

Source: Watson, K.S., EPA Technology Transfer Program;   March, 1973

TABLE 79:   MONTHLY AVERAGES OF THE INFLUENT TO SOUTH
EDMESTON TREATMENT FACILITY

	Avg. Max.	Avg. Min.	Avg.	Avg. lbs/day
BOD mg/l	552	408	485	440
COD mg/l	1033	955	1010	916
Suspended solids	208	128	162	147
Flow gal/day	142,758	73,880	108,778	

Source: Watson, K.S., EPA Technology Transfer Program; March, 1973

After biological treatment in the lagoon, the effluent enters a tank which
is divided into two compartments and passes through the clarifier section,
with a 2.5 hour holding capacity, which is served by a sludge raking
flight. The remainder of the tank consists of a compartment for holding
sludge which is removed from the effluent by the clarifier. This sludge is
returned to the aeration basin through the use of a sludge sump and pump
located adjacent to the tank. The effluent is chlorinated at the flow
measuring device to comply with state requirements because it also carries
the sanitary discharge from the plant.

The treatment facility was designed for a BOD loading of 560 pounds per
day and an effluent volume of 172,000 gpd. The design called for removing 85% of the BOD and 90% of the suspended solids. As shown by
figures listed in Table 79, the treatment facilities are at present somewhat
underloaded. The removal of BOD is averaging about 95% (Table 80).
The removal of suspended solids is somewhat less efficient perhaps because
the concentration is not high to start with. In both cases effluent concentrations fall well within state specifications.

As a result of operating experience, the aerator, which was sized to provide 1.5 pounds of oxygen per pound of BOD, is controlled by a timer.
It is in operation about 66% of the time which keeps the dissolved oxygen
content of the basin between 5 and 6 mg./l. Such an operating mode
provides expansion capacity in the treatment facilities and tends to improve
sludge settleability over higher oxygen concentrations being maintained
in the basin.

Provisions have been made in this system for adequate sampling and analyses

to control its operation. The treatment plant influent is sampled in a
representative manner through a valve on the discharge side of the pump
delivering all the wastewater to the system. As the effluent leaves the
clarifier, the flow is recorded and totalized using a Kenneson nozzle.

TABLE 80:　MONTHLY AVERAGES OF THE EFFLUENT OF THE SOUTH
EDMESTON TREATMENT FACILITY

	Avg. Max.	Avg. Min.	Avg.	Avg. lbs/day
BOD mg/1	31.8	11.5	20.8	18.9
COD mg/1	68	25	47.9	43.5
Suspended solids	56	17	32	29
Flow gal/day	142,758	73,880	108,778	

Source:　Watson, K.S., EPA Technology Transfer Program; March, 1973

The laboratory which handles the wastewater analyses for this plant and
another located nearby is located within the process plant. Tests which
are run include BOD, COD, suspended solids, total solids and volatile
suspended and total solids. Operational tests for controlling the treatment
system consist of dissolved oxygen, mixed liquor suspended solids, settleable solids and chlorine residual.

CHAMPAIGN, ILLINOIS

Kraft Foods has had a margarine and salad dressings plant in operation in
Champaign since the early 1960 period. In 1968 the decision was reached
to make a major expansion at the Champaign plant.

A portion of the facility is in general the original oil plant in which margarine, salad dressings and oil products are manufactured. The second
portion of the facility is new, having been placed in operation over a
year ago and in which macaroni type products, process cheeses (slices and
Velveeta type) and natural cheeses in consumer size packages are produced.
A plant immediately behind the Kraft plant is an edible oil refinery operated
by the HumKo Operation of the Corporation. It should be noted that the
close proximity of these facilities to extensive residential areas means that
all phases of environmental control must be carefully practiced.

In planning for the plant expansion, early consideration was given to
proper handling of the liquid wastes problem for the expanded facility.
As is the corporation's usual approach, it was proposed to the Urbana-
Champaign Sanitary District that the district provide sewerage service
for the expanded plant at Kraft's expense since the wastes to be discharged
would be completely compatible with district wastewater and be degraded
in its professionally operated treatment system.

After considerable negotiation, the district stood fast on its position that
it could accept the hydraulic load but the plant would need to provide
treatment facilities to meet the specification of: 200 mg./l. of BOD,
200 mg./l. of suspended solids and 100 mg./l. of fats, oils and greases
covered in a proposed ordinance.

Since business considerations dictated that the production capacities and
mix of products already mentioned should be located at Champaign, the
only course open was to provide the same type of biological treatment
facilities as those operated by the district. These facilities, of course,
must be operated from now on.

As a result of the configuration of the site, it was necessary to separate
the primary facilities from the secondary. The design basis for the plant
is shown in Table 81. It will be noted that the facility has been designed
for a projected 1980 load.

TABLE 81:   BASIS OF DESIGN FOR 1980 LOAD

Parameters	Design
Average flow, gpd	500,000
Average flow, gpm	350
BOD, mg./l.	3,640
BOD Load, lbs./day	15,000
Suspended solids, mg./l.	685
Suspended solids, lbs./day	2,850
Grease, mg./l.	3,140
Grease, lbs./day	13,000

Source: Watson, K.S., EPA Technology Transfer Program; March, 1973

The wastes from the cheese and oil production facilities at Champaign are
collected in lift stations and pumped to a surge tank. The items of equipment used in the treatment system are summarized in Table 82.

TABLE 82:   ITEMS OF EQUIPMENT IN THE TREATMENT SYSTEM

PRIMARY PLANT

Lift Stations

    Cheese   - 2 - 225 gpm - 10.0 H.P. Pumps
    Oil Plant - 2 - 250 gpm - 7.5 H.P. Pumps

Surge Tank

    1 - 30'D x 20'H - 80,000 gal. with 10 H.P. agitator
                            and sludge rake
    Minimum Detention Time - 1.5 hours (avg. flow and 6' level)
    Max. Detention Time   - 4.5 hours (avg. flow - 18' level)

Flotation Clarifier

    1 - 39'D x 10'H - 76,000 gal. with sludge rakes and surface
          skimmer and recycle pressurization system.
    At 50% recycle and avg. flow - surface settling rate = 625 gpd/ft^2
    Weir overflow rate - 2450 gpd/lineal ft.; detention time = 2.45 hrs.

Primary Sludge Storage Tank

    1 - 30'D x 9'-6"H Covered Tank - 42,500 gal.

Grease Storage Tank

    1 - 14'D x 29'H Cone Bottom, covered, heated tank - 18,000 gal.

SECONDARY PLANT

Aeration Basin

    Rectangular Basin 309 x 149 x 9' Deep
    Operation Volume - 2,270,000 Gal.
    Detention Time @ avg. flow - 4.5 days

    Dorr-Inka Aeration system using 4-75 HP blowers @ 8,000 scfm each

Final Clarifier

    1 - 49'D x 8½'H tank - 100,000 gal. with sludge rakes and
                             surface skimmers
    At avg. flow - Surface settling rate = 270 gpd/ft^2
    Weir overflow rate = 325 gpd/lineal ft
    Detention time = 4.8 hours

Aerobic Digestor

    1 - 50'D x 27'H tank - 368,000 gal. with
    3 - 50 H.P. Blowers - 700 scfm @ 7.5 Psig each

Sludge Lagoons (located at Sanitary District's plant site)

    2 - 1 acre surface x 8'D earthen tanks

Source: Watson, K.S., EPA Technology Transfer Program; March, 1973

The next step in the process is the flotation clarifier. Grease skimmed from the surface of this unit is conveyed to the grease tank already mentioned. The sludge removed from the clarifier is passed on to the aeration basin to avoid the need for primary sludge handling facilities. Next, the effluent flows to the aeration basin, the first step in the secondary treatment portion of the system. The discharge from the aeration basin then passes through the final clarifier and into the district system.

Arrangements have been made for collecting samples at various points throughout the system. An automatic, composite sampler is located on the final discharge so a continuous record can be developed on the load being contributed to the district. Laboratory control results for the operation of these treatment facilities are run in the quality control laboratory.

Table 83 shows the quality of the effluent being discharged into the sanitary district system at Champaign from the middle of December 1972 until the end of January 1973 based on daily analyses. Unfortunately the plant decided to make some changes in its sampling and analytical regime in October and had not again started running BOD analyses on the influent to the plant during the period under consideration. The monthly maximum, minimum and mean BOD results on the influent for a three month period ending with October were respectively 5,268, 1,113 and 3,223 mg./l.

TABLE 83:   THE EFFLUENT BEING DISCHARGED INTO THE SYSTEM

	FLOW mgd	BOD mg/1			COD mg/1	SS mg/1
	Mean	Max	Min	Mean	Mean	Mean
Dec. 15-21, 1972						
Influent	0.283	-	-	-	4397	1036
Effluent		90	45	54	157	61
% Removal					96.4	94.1
Dec. 22-28						
Influent	0.210	-	-	-	5176	1200
Effluent		221	74	125	376	93
% Removal					92.7	92.2
Jan. 1-7, 1973						
Influent	0.233	-	-	-	5157	1244
Effluent		113	32	73	254	92
% Removal					95.1	92.6
Jan. 8-14						
Influent	0.269	-	-	-	7471	1807
Effluent		82	32	58	192	74
% Removal					97.4	95.9
Jan. 15-21						
Influent	0.262	-	-	-	5434	1210
Effluent		214	54	127	337	97
% Removal					93.8	92.0
Jan. 22-28						
Influent	0.280	-	-	-	6225	1368
Effluent		142	99	117	438	184
% Removal					93.0	86.5

Source: Watson, K.S., EPA Technology Transfer Program; March, 1973

Thus, if the influent BOD had remained in the same general range for the six week period under consideration, BOD reductions would have ranged between roughly 90 and 98%. It is apparent from Table 83 that the reduction in suspended solids resulting from passing the effluent through the plant ranged between 85.6 and 95.9%. Further, the COD reduction ranged between 93 and 97.4%.

Since the more work that is done on the degrading of edible fats, oils and greases in a plant effluent the clearer it becomes that these types of materials are rather readily degradable in biological treatment systems, this discussion will be concluded by a survey of this situation at the Champaign plant.

Table 84 summarizes the concentration of fats, oils and greases on a monthly basis in the influent and effluent of the treatment plant for the last six months of 1972. The reductions from influent to effluent across the system are accomplished by a combination of the primary and secondary treatment processes in use. It is of interest to note that an average of 97.4% of the fats, oils and greases were removed or degraded by the treatment system described.

TABLE 84: REMOVAL OF FATS, OILS AND GREASES BY THE CHAMPAIGN TREATMENT PLANT, 1972

MONTH	INFLUENT mg/1			EFFLUENT mg/1			% REMOVAL $\% = (100) \dfrac{\text{Inf.(Mean)} - \text{Eff.(Mean)}}{\text{Inf. (Mean)}}$
	Max.	Min.	Mean	Max.	Min.	Mean	
July	2,780	161	1,376	98	29	47	96.6
Aug.	9,023	494	1,945	57	25	35	98.2
Sept.	1,966	416	1,176	50	27	39	96.7
Oct.	2,845	181	1,134	111	15	41	96.4
Nov.	4,002	215	1,058	108	6	21	98.0
Dec.	2,314	243	1,158	47	4	17	98.5
TOTAL	22,930	1,710	7,847	472	106	200	584.4
AVERAGE	3,822	285	1,308	79	18	33	97.4

Source: Watson, K.S., EPA Technology Transfer Program; March, 1973

CHEMUNG, ILLINOIS

The Dean plant at Chemung (13) is a fluid milk plant processing about 1.1 million pounds of milk per day. This plant processes and bottles a complete

line of fresh dairy products including soft serve ice milk mixes. Cultured
products, sour cream, buttermilk, and yogurt are also manufactured at
this location. Cottage cheese was produced at this plant until April, 1972.

Waste control and wastewater treatment have been a part of this plant
operation since 1950. This has been necessary because Chemung is a very
small community of only a few homes and does not have a municipal waste
disposal plant. Throughout the years improvements for waste reduction and
segregation have been made both in the plant and in the treatment system.
The following examples are some of the practices used to recover and
segregate waste inside the plant:

1. Fresh pasteurized product losses are recovered for use in ice
   cream mix.

2. Product losses which cannot be reused are segregated for
   animal feed.

3. Whey from cottage cheese manufacture was recovered.

4. Cottage cheese rinsewater was disposed of through sprinkler
   irrigation.

5. Uncontaminated water is segregated for cooling tower treatment and the discharge bypasses the treatment plant.

The unsegregated waste is discharged to the waste treatment plant and
consists of approximately 80,000 gpd and 1,150 pounds of BOD_5 per day.
Wastewater coefficients for the unsegregated waste load are approximately
0.6 lb. wastewater per lb. of milk and 1.0 lb. BOD_5 per 1,000 lbs. of
milk. Table 85 shows the monthly averages for these characteristics during
1971 and 1972. The daily averages and the daily maximum and minimum
are also given in this table.

The waste treatment facility consists of an activated sludge system followed
by two lagoons. A flow scheme for this system is given in Figure 78. The
influent enters an aerated waste holding tank (A) where the flow and BOD_5
is partially equalized. Nitrogen is also added to the influent at this point.

The waste then goes into two activated sludge aeration tanks (B) which
are operated in series. The retention time in these tanks is approximately
24 hours and a BOD_5 reduction of about 75% is obtained. The next step
is two gravity clarifiers (C) which operate in parallel with approximately
80,000 gpd effluent overflow and 160,000 gpd return sludge underflow.
Activated sludge effluent receives additional treatment in two aerated
lagoons (D) operated in series with 20 days retention time to obtain an

additional BOD_5 reduction of about 70%. The total BOD_5 reduction is 90 to 95%. A lagoon settling zone (E) of four days retention time achieves a SS reduction of approximately 85%. The lagoon effluent is chlorinated (F) and mixed with the segregated cooling water prior to discharge to the stream. An aerobic sludge digester (G) is used to further reduce BOD_5 and concentrate SS in waste activated sludge. The digested sludge is applied to an approved 21 acre irrigation site.

TABLE 85:   DAIRY PROCESS WASTEWATER CHARACTERISTICS FOR DEAN FOODS COMPANY, CHEMUNG, ILLINOIS

MONTH	FLOW gpd	BOD_5 mg/l	BOD_5 POUNDS PER DAY	pH	# WASTE WATER PER # MILK	# BOD_5 PER 1000 # MILK
1-71	88,000	1494	1096	7.49	0.56	0.83
2-71	93,000	2041	1583	7.87	0.55	1.13
3-71	83,000	2070	1433	7.79	0.51	1.06
4-71	91,000	1575	1195	7.02	0.59	0.93
5-71	92,000	1476	1132	7.72	0.62	0.91
6-71	98,000	2117	1730	6.87	0.75	1.58
7-71	89,000	1900	1410	6.80	0.63	1.23
8-71	98,000	1818	1486	6.60	0.72	1.30
9-71	101,000	1135	956	6.65	0.73	0.80
10-71	104,000	1609	1396	7.67	0.73	1.25
11-71	80,000	1215	811	7.90	0.57	0.69
12-71	80,000	2470	1648	7.40	0.55	1.36
1-72	64,000	1520	811	7.14	0.43	0.66
2-72	63,000	1638	860	6.52	0.49	0.80
3-72	61,000	1450	738	6.32	0.42	0.59
4-72	63,000	1430	751	7.14	0.48	0.68
5-72	63,000	1368	719	6.85	0.51	0.69
6-72	69,000	1618	930	7.00	0.59	0.95
7-72	81,000	1854	1252	6.82	0.59	1.25
8-72	90,000	2242	1683	7.30	0.68	1.53
9-72	82,000	1570	1074	6.75	0.62	0.97
10-72	70,000	1207	705	7.10	0.50	0.61
11-72	76,000	1604	1017	8.24	0.53	0.91
12-72	75,000	1962	1227	7.33	0.55	1.18
Daily Avg.	81,400	1712	1156	7.16	0.58	0.98
Daily Max.	140,000	3400	2667	9.10	0.75	1.58
Daily Min.	23,000	817	670	5.60	0.42	0.61

Estimated SS = 300 mg/l or 200 pounds per day

Source: Watson, K.S., EPA Technology Transfer Program; March, 1973

FIGURE 78:   WASTE TREATMENT FLOW SCHEME

A.   Raw waste equalization tank
B.   Activated sludge aeration tanks
C.   Clarifiers
D.   Aerated lagoons
E.   Settling lagoon
F.   Chlorinator
G.   Sludge aerobic digester

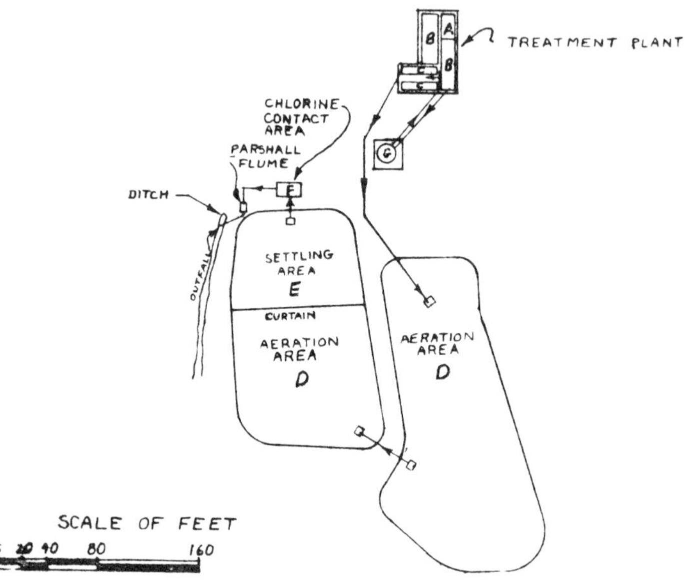

Source: Watson, K.S., EPA Technology Transfer Program; March, 1973

The effluent characteristics from the various steps in the operation are shown in Table 86. These figures are monthly averages for 1971 and 1972 and again the daily averages and maximum and minimums are given. The final effluent consists of approximately 80,000 gpd chlorinated lagoon effluent and 200,000 gpd cooling water discharge. The combined effluent contains approximately 30 to 60 mg./l. BOD_5, 35 mg./l. SS, 25 mg./l. phosphate, 5 mg./l. ammonia and 0.5 mg./l. nitrate.

The major components of the waste treatment complex were initially installed as follows:

1950 — One aeration tank and one clarifier
1961 — Second aeration tank and clarifier
1967 — Sludge digester and irrigation site
1968 — Lagoons and chlorinator

The replacement cost of these components is estimated at over $500,000. Annual operating cost is estimated at $50,000 including personnel.

Since January 1972 a technically educated operator has been employed to conduct waste sampling and analyses, interpret results, and operate the treatment plant in a scientific manner. Prior to that time the responsibility was divided between the quality control laboratory and maintenance department.

TABLE 86:   EFFLUENT CHARACTERISTICS FOR DEAN FOODS COMPANY, CHEMUNG, ILLINOIS

MONTH	ACT. SLUDGE EFF.		LAGOON EFF.		FINAL		EFF.
	BOD_5 mg/l	SS mg/l	BOD_5 mg/l	SS mg/l	BOD_5 mg/l	SS mg/l	DO mg/l
1-71	964	680	258	111	77	14	7.1
2-71	1001	910	247	30	148	79	
3-71	602	764	201	45	98	29	7.3
4-71	815	585	180	30	64	43	5.1
5-71	160	40	795	634	114	83	7.1
6-71	365	595	120	40	56	35	7.3
7-71	136	268	58	25	44	14	8.1
8-71	115	342	65	25	26	15	7.5
9-71	581	578	68	21	22	14	7.0
10-71	926	499	70	30	32	55	7.7
11-71	883	953	65	30	19	26	8.1
12-71	581		84		44	12	6.6
1-72	875	1246	135	78	49	6	8.0
2-72	540	303	129	49	103	36	12.7
3-72	744	648	109	63	15	26	7.2
4-72	46	31	36	123	35	28	7.4
5-72	129	170	137	78	23	28	8.0
6-72	114	102	100	65	33	44	9.1
7-72	825	788	84	36	8	25	6.3
8-72	325	221	104	48	45	22	8.1
9-72	165	848	80	86	11	56	7.6
10-72	82	107	69	81	6	55	6.9
11-72	132	298	51	42	35	59	7.9
12-72	172	349			23	40	7.8
Daily Ave.	470	492	139	80	47	35	7.6
Daily Max.	2100	1300	2200	1200	260	303	16.4
Daily Min.	20	30	40	10	1	4	3.5

Source: Watson, K.S., EPA Technology Transfer Program; March, 1973

The final effluent has had little impact on the receiving stream as shown by the upstream and downstream data in Table 87. The final effluent of approximately 280,000 gpd is discharged to a stream with a ten year flow of around 890,000 gpd. Upstream and downstream BOD_5 and DO have been nearly identical.

TABLE 87:   WATER QUALITY - RECEIVING STREAM FOR DEAN FOODS COMPANY, CHEMUNG, ILLINOIS

MONTH	UPSTREAM		DOWNSTREAM	
	BOD_5 mg/l	DO mg/l	BOD_5 mg/l	DO mg/l
1-72	5	13.6	5	13.5
2-72	10	14.4	11	14.5
3-72	14	10.9	12	10.2
4-72	8	12.2	6	12.2
5-72	4	14.6	2	14.7
6-72	4	10.1	5	11.9
7-72	9	11.2	11	11.7
8-72	3	11.3	3	12.5
9-72	3	11.0	6	11.3
10-72	3	12.0	3	12.1
11-72	3	13.5	1	14.0
12-72	1	14.1	0	14.3
Avg.	6	12.4	5	12.7
Daily Max.	30	16.0	30	16.6
Daily Min.	0	8.0	0	9.6

Source: Watson, K.S., EPA Technology Transfer Program; March, 1973

KENT, ILLINOIS

The Kent Cheese Company specializes in the production of ricotta, parmesan, romano, and mozzarella cheese. During the study (66) an average of 13,000 pounds of cheese were produced per day (excluding Sunday) producing approximately 17,000 gallons per day of wastewater with a BOD of 270 pounds per day and a suspended solids loading of 85 pounds per day. Sources of wastewaters were from the rinses and washes associated with milk storage, transmission lines, vats and pasteurizer. Whey was collected and transported to another site for recovery.

The wastewater treatment system consisted of two equal volume aerated lagoons in series. Each 12 foot deep lagoon holds 955,000 gallons providing a detention time of 56 days based on design flow. The first lagoon was provided with thirteen 6 foot long 18 inch diameter Helixors (Polcon Corp.) arranged in a pattern along the flat portion of the lagoon bottom. Three additional aerators are arranged in a triangular pattern near the inlet end. Two 240 scfm Gardner Denver rotary blowers provided the air supply for the lagoons. Water surface elevation in the lagoons was controlled by placement of a 4 inch cast iron riser pipe at a fixed elevation to maintain a 12 foot water depth.

The aerators were of the submerged air lift type insuring operation throughout the year. Oxygen transfer studies conducted during a one year study indicated that standard transfer rates ranging from 2.2 to 4.1 lbs./hp. hr. were achievable. Low mixing velocities were observed with the diffuser pattern and basin geometry employed in this lagoon. Very light accumulations of sludge (less than 2 inches) were noted after one year.

Results of the lagoon performance based on BOD appear in Table 88. Wastewater temperature had the greatest effect on process performance. An overall average removal of 97.3% produced an average BOD concentration of 52 mg./l., ranging from 155 mg./l. to 12 mg./l. during the one year study. Poorest performance occurred during the winter months (Jan. and Feb.) but high oxygen uptake rates occurring in April through mid July produced zero dissolved oxygen concentration in the primary lagoon causing odor and sludge rising problems. This unusually high activity has been attributed to rapid anaerobic decomposition of benthal solids accumulated over the cold winter months.

During the one year study approximately 65% of the nitrogen was removed. Highest nitrate production occurred in warm summer months. Effluent total nitrogen concentrations ranged from 0.2 to 10.6 mg./l. averaging 3.8 milligrams per liter. Total phosphorus removals of approximately 50% were observed resulting in an average total phosphorus concentration of 21.6 milligrams per liter.

TABLE 88:   PERFORMANCE OF TWO STAGE LAGOONS, KENT CHEESE CO.

Quarter 1971	Flow (gal/d)	Influent		Prim.Lagoon Eff.		Sec.Lagoon Eff.		Overall Removal		Temp.
		BOD (mg/1)	SS (mg/1)	BOD (mg/1)	SS (mg/1)	BOD (mg/1)	SS (mg/1)	BOD (mg/1)	SS (mg/1)	°C
Jan–Mar	14,900	1,940	658	224	403	106	155	94.5	73.4	1
Apr–June	17,300	2,040	600	204	477	61	119	97.0	80.1	17.8
July–Sept.	20,000	1,530	547	122	239	21	43	98.5	92.1	21.7
Oct–Dec.	15,200	2,100	595	274	445	31	111	98.5	81.3	9.5
Average	16,900	1,910	602	209	395	52	108	97.3	82.0	--

Source: Boyle, W.C. and Polkowski, L.B., EPA Technology Transfer Program; March, 1973

Total capital investment for this plant in 1970 was $49,500 or $2,900 per 1,000 gallons of design flow. This is substantially higher than the 1963 costs estimated for aerated lagoons for cheese processes. Over the first year of the study the operation and maintenance costs were approximately $6,000 resulting in unit costs of approximately $0.13 per pound of BOD and $2.05 per 1,000 gallons.

DE PERE, WISCONSIN

The Eiler Cheese Company processes 30,000 pounds of milk per day into 3,000 pounds of American cheddar and Colby cheese (67, 68). The treatment plant was designed to handle 3,000 gallons per day of wastewater, with a maximum of 5,400 gallons per day, and a BOD load of 90 pounds per day (2,000 mg./l.). About half of the milk handled is received in cans, the washings from which constitute the major source of wastewater. The balance of wastewater results from whey washing. Whey is separated and hauled to local farmers as a feed.

The wastewater treatment facility consists of three septic tanks connected in series followed by a four stage rotating disc system, a clarifier, a chlorinator, and a polishing lagoon (Figure 79). The septic tanks provide a flow equalization, clarification of raw wastewater and digestion of recycled biological sludge. The three cells have volumes of 6,450 gallons, 5,040 gallons, and 1,400 gallons respectively providing detention times ranging from 2.5 to 4.3 days. The rotating biological filters furnished by Autotrol Corporation were located in a vault 12 feet below ground level. The BIO-DISC (Autotrol Corp.) system consisted of feed chamber, four stage BIO-DISC unit, and clarifier.

FIGURE 79: EILER CHEESE COMPANY BIO-DISC TREATMENT PLANT

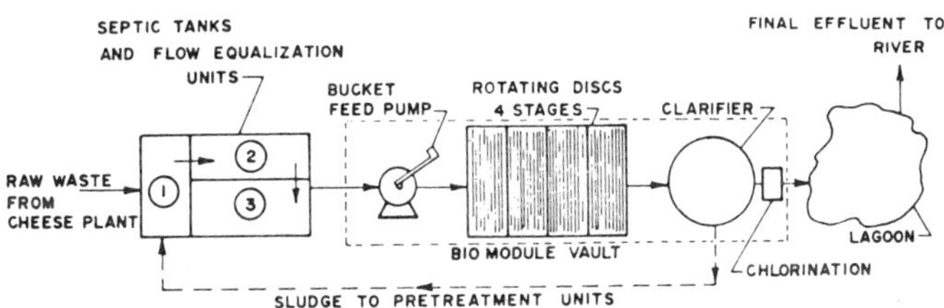

Source: Boyle, W.C. and Polkowski, L.B., EPA Technology Transfer Program; March, 1973

TABLE 89:   PERFORMANCE OF SEPTIC TANK, BIO-DISC, EILER CHEESE PLANT

Date	Flow	Raw Waste			Septic Tank Eff.			BIO-DISC Eff.			Polishing Pond
		BOD	SS	pH	BOD	SS	pH	BOD	SS	pH	BOD
1971	(gal/day)	(mg/1)	(mg/1)		(mg/1)	(mg/1)		(mg/1)	(mg/1)		(mg/1)
Apr.19	3676	840	360	7.6	863	240	6.7	40	80	7.7	17
20	3411	720	280	7.6	810	320	6.6	32	20	7.6	27
21	3625	780	220	7.6	640	220	6.7	21	80	7.8	32
22	4405	1700	340	7.3	675	180	6.7	29	20	7.7	22
23	4703	1240	640	7.1	955	280	6.7	54	40	7.4	53
24	1540	1100	220	7.1	1060	160	6.7	65	40	7.5	53
25	761	705	140	7.2	870	120	6.5	53	20	7.6	50
26	3477	840	460	7.1	1000	100	6.6	30	60	7.5	48
27	3030	1540	240	7.2	970	240	6.6	48	40	7.6	46
28	3825	1150	240	7.0	1120	200	6.6	48	60	7.6	17
Avg.	3245	1062	314	7.3	852	206	6.7	41	46	7.6	37

Source: Boyle, W.C. and Polkowski, L.B., EPA Technology Transfer Program; March, 1973

The feed chamber is provided with a bucket feeder attached to the BIO-DISC shaft, thereby providing a constant feed to the biological unit. Each of the four BIO-DISC stages contains 22 molded polystyrene discs, 10 feet in diameter, providing a total area of 13,800 square feet. The discs are rotated at a speed of 2 rpm (peripheral velocity of 62 feet per minute). Each stage provides a hydraulic detention time of 1.5 to 2.0 hours. The clarifier provides for overflow rates of 1,800 to 2,700 gallons per day per square foot with sludge removal being provided by sludge scoops rotating at 4.5 rph. Sludge flows by gravity back to septic tank cell #1. A 4.5 foot deep polishing lagoon with a 30 day detention time is provided to "polish" final effluent from the BIO-DISC.

Results of the performance of this treatment system for a 12 day period is presented in Table 89. It is apparent from examining this table that the septic tank did not provide the treatment expected. It was found that pH values dropped to values of 5.5 in cell #2 resulting in inhibition of methane bacteria. It should be noted, however, that recovery of pH did occur on the rotating biological filters and that excellent BOD removals were still achieved. BOD removal exceeded 95% on the BIO-DISC throughout the test period of 12 days. It should also be noted that the "polishing" value of 30 day lagoons is questionable.

Burying of the BIO-DISC vault provided substantial insulation against cold winter temperatures. Minimum temperatures of the BIO-DISC mixed liquor never fell below 40°F. even though ambient temperatures as low as -30°F. were recorded. Several instances of shock loading were recorded during the 10 months of recorded data. Increased production resulting in peak discharges, raw milk spillage, and a whey dump were all experienced. The BIO-DISC unit continued to provide 60 to 75% treatment with full recovery after two to three days.

Capital investment for this process was $35,000 or $6,500 per 1,000 gallons of design capacity. These figures are estimated, based on current biomodule cost projection. Operation and maintenance are minimal. The operating power for the disc drive was 0.5 hp. and the clarifier scraper was driven by a 1/6 hp. motor resulting in an annual power cost of $100 based on $80 per horsepower year. Miscellaneous costs for septic tank cleanout and approximately one hour per month for servicing would cost about $200. Total annual operating costs, then, would be approximately $300 or $0.035 per pound of BOD.

AFOLKEY, ILLINOIS

The Afolkey Coop. Cheese Co. produces approximately 8,000 pounds per day of Italian pizza cheese (69). Wastewater flows vary from 3,600 to

9,000 gallons per day with an average BOD concentration of 3,500 mg./l.
Wastewaters are generated from milk spillage and equipment washup. All
milk is received in bulk trucks. Cooling water is separately discharged and
whey is hauled to local farmers for livestock feed.

The wastewater treatment facility consists of an existing septic tank,
appropriately modified, an aerated lagoon, a quiescent lagoon, and a
sand filter (Figure 80). The existing 28,200 gallon three celled septic
tank providing an average hydraulic detention time of 3.1 days (maximum
flow) was modified by the addition of paddle mixers placed within the
first two equal sized tanks.

The slow speed mixers were operated through a timer which regulated
mixing for 15 minutes out of every two hours during times when the plant
flow was off. This agitation provides intimate contact of anaerobic
organisms with the raw wastewater. During periods of process operation,
the mixers are shut off so as to allow maintenance of high concentrations
of active biomass in the first two cells. The third cell provides for quiescent
settling prior to discharge to the aerated lagoon.

FIGURE 80:   AFOLKEY CHEESE COMPANY WASTEWATER TREATMENT
PROCESS

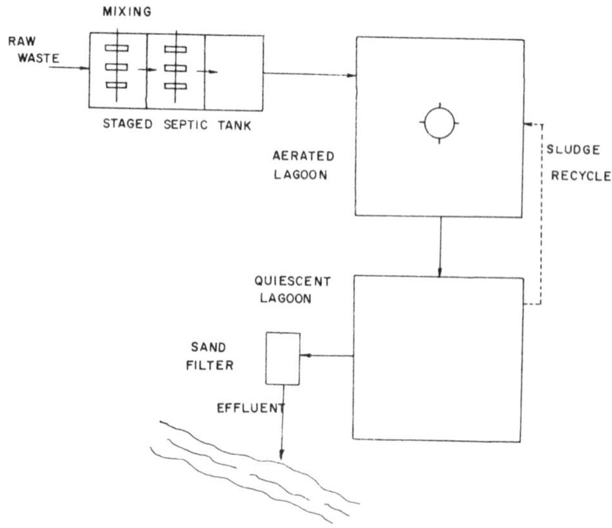

Source: Boyle, W.C. and Polkowski, L.B., EPA Technology Transfer
Program; March, 1973

The 10 foot deep aerated lagoon with a volume of 40,500 gallons provides a hydraulic detention time of 4.5 days at maximum flow. Aeration is provided by a 10 horsepower floating aerator located at the lagoon center. Sludge, pumped from the quiescent lagoon, may be returned to the aerated lagoon to provide high biomass concentrations if required. Although originally uncovered, the aerated cell is now covered to reduce icing problems during the winter season.

The 10 foot deep quiescent lagoon serves to provide for settling and additional stabilization. This lagoon has a volume of 20,900 gallons providing 2.3 days of detention time at maximum flow. A baffled overflow weir is provided along the entire width at one end for effluent discharge to the sand filter. The tank is equipped with a hopper bottom to allow sludge collection and removal by pumping. This tank was initially covered to eliminate excessive algal growths during the summer.

The sand filter was added to provide some means of effluent polishing. It consists of a 4 ft. by 8 ft. box approximately 12 inches deep. Approximately 6 inches of "filter sand" is underlain by coarse sand and pea gravel. Underdrain tiles carry the filtered wastewater by gravity to the receiving stream. The wastewater is applied at a maximum rate of 11.7 gallons per hour per square foot (0.2 gallon per minute per square foot). The filter is manually cleaned when the headloss exceeds approximately four or five feet. A small layer of old sand and sludge is removed and clean sand is added in this process. Cleaning is required approximately every three or four days although shorter periods of cleaning are required during periods of high solids overflow.

Data on the performance of this plant are sketchy; the results of two surveys are presented in Table 90. The April 4, 1972, data were obtained when the sand filter was not in use. Results of septic tank performance are also presented in Table 89. These data were collected prior to installation of the slow speed mixers in cells #1 and #2.

There is every indication from the data collected to date that the system will produce a highly stabilized effluent well below the 30 mg./l. BOD currently required. It should be noted, however, that this performance is based upon flows considerably less than maximum. The best estimate of flow during the surveys reported was 3,600 gallons per day, thereby resulting in detention time in the treatment units as follows:

Septic tank	7.8 days
Aerated lagoon	11.2 days
Quiescent lagoon	5.8 days
Sand filter	4.7 gallons per hour per square foot

TABLE 90:   PERFORMANCE OF SEPTIC TANK, LAGOON SYSTEM, AFOLKEY COOP. CHEESE CO.

April 4, 1972

Est. Flow = 3600 gal/day
Est. BOD = 3500 mg/l

Final Effluent - No sand filtration
BOD = 15 mg/l
TSS = 16 mg/l
pH = 8.1
Nitrate N = 4.8 mg/l

October 20, 1972

Est. Flow = 3600 gal/d
Est. BOD = 3500 mg/l

Final Effluent - Sand Filter
BOD = 6 mg/l
TSS = 22 mg/l
pH = 7.6
NH_3-N = 0.13 mg/l
Kjeldahl-N = 2.91 mg/l
Nitrate-N = 14.5 mg/l
COD = 27 mg/l
Total Phosphorus-P = 16.3 mg/l

Septic Tank Performance (Prior to Mixing)

	BOD mg/l	T.S.S. mg/l	pH
Dec. 1968			
Cell 3	435	--	6.1
Jan. 1969			
Cell 1	2070	930	5.4
Cell 2	1245	440	5.6
Cell 3	760	370	5.6
Feb. 1969			
Cell 3	815	560	---

Source: Boyle, W.C. and Polkowski, L.B., EPA Technology Transfer Program; March, 1973

The capital investment in this plant in 1971 was approximately $25,000 or $2,780 per 1,000 gallons of design capacity. Operation and maintenance costs include approximately $1,440 per year for power, $150 per year for septic tank pumping two or three times per year, and $10 per day for plant maintenance. That amounts to approximately $4,200 per year or unit costs of $1.27 per 1,000 gallons or $0.004 per pound of BOD based on maximum design flow (9,000 gallons per day at 3,500 mg./l.).

RALEIGH, NORTH CAROLINA

A detailed study of a typical North Carolina multiproduct dairy plant has been made by R.E. Carawan, V.A. Jones and A.P. Hansen of the Department of Food Science at North Carolina State University, Raleigh, North Carolina (14). Products processed by the plant included a complete line of fluid products (82% of total production), by-products including cottage cheese (7%) and frozen products including novelties (11%). Maximum rate of milk usage was 40,000 gallons per day.

Water use totaled 325.5 gallons per 1,000 pounds of total product. The processing area was the largest user requiring 311.4 gallons per 1,000 lbs. of total product. Water use by product grouping ranged from 259.6 gallons per 1,000 pounds for fluid products to 1,397 gallons per 1,000 pounds for frozen products. The largest water use by operation was for utilities which required 38.8% of the total water. The case washer was identified as using 8.4% of the processing water and suggestions were made for reduction in this water use. Wide variation was noted in production and water use from day to day.

The effluent from the processing area was sampled at two locations. Effluent discharged from the processing area averaged 73.67% of the water use. The process area effluent discharge was 300 gallons per 1,000 pounds of milk received. Waste parameters were identified and included BOD, 7.34 lbs./1,000 lbs. product (2,257 mg./l.); TSS, 3.59 lbs./1,000 lbs. product (1,104 mg./l.); and fat, 2.34 lbs./1,000 lbs. product (720 mg./l.). The frozen products area contributed 54% of the BOD and 69% of the fat in the processing area waste stream. Wide variations were noted in the waste characteristics from day to day.

Effluent temperatures were monitored and the fluid and by-product drain averaged 95° to 100°F. while the frozen products drain averaged 85° to 90°F.

Municipal charges for dairy plants were established and total cost for water, sewer and surcharge for cottage cheese production were 0.9¢ per pound. A dairy plant processing 500,000 pounds of product each day will find itself paying an approximate $6,000 monthly municipal bill using average values obtained in this study (14): An average BOD/COD ratio was found to be 0.64; the BOD/TOC ratio varied from 0.74 to 2.22.

Plant milk and fat loss records were compared with the results of effluent analyses. Losses based on BOD and fat test data exceeded plant loss records. A program to help dairies reduce their water use and waste generation was presented (14). This program stressed the need for minimizing water use and waste discharge to achieve economic and environmental goals.

FUTURE   TRENDS

As municipalities increase the practice of charging the industry for its
waste on a compositional basis and/or impose legal limits on waste composition going to municipal treatment facilities, the industry will take a more
active role in fluid waste control. An active nationwide educational program is needed to influence attitudes, to provide a sound technological
basis for controlling the pollution problems of the dairy industry of the
future.

Using projected changes in plant numbers and production per plant, the
total projected production of various dairy products between 1970 and 1985
has been projected. See Figure 1 (page 2). The values predicted for 1977
are in general agreement with those projected in the Cost of Clean Waters
Waste Profile No. 9 (3).

The projection presented in this report assumes that the trends of the past
decade will continue into the future. In this regard, it is important to
point out that those segments of the dairy food industry that are potentially
the greatest environmental polluters are also those segments of the industry
showing the most growth potential. These subindustries, cottage cheese,
cheese, and ice cream are the only segments of the overall dairy industry
to show increases in per capita consumption for the past decade.

As plants become larger in the future, there will be greater utilization of
mechanization and automation in all phases of dairy food plant operations.
Waste Profile No. 9 in 1967 (3) adequately outlined the types of dairy
food processes and indicated the differences in old, typical and advanced
technology. The larger plants today and those of the future will rely upon
advanced technology, and automation will be the center of this technology.
Key advanced technologies applicable to all types of dairy food plants that

have significance to dairy plant wasteloads include:

    (a) Milk receiving in tank trucks with automated rinsing
         and cleaning of the tankers at the dairy plant.

    (b) Automated dairy food processing operations processing
         milk at rates up to 100,000 lbs./hr., including the
         use of Cleaned in Place separators and clarifiers that
         automatically discharge sludge every 15 to 30 minutes.

    (c) Automated circulation cleaning, using liquid detergents
         and chemical sanitizing agents on a controlled basis.

    (d) High speed filling and packaging operations.

    (e) Automated materials handling, using conveyors, casers
         and stackers requiring significant quantities of lubri-
         cants.

Theoretically, advanced technology should reduce water volumes used in dairy food plants and provide for reduced wasteloads. Waste Profile No. 9 indicated that BOD waste coefficients (pounds BOD/pound of product) became less with the use of advanced technology, with plant size having no effect on waste coefficients. However, the use of advanced technology in itself is no guarantee of low waste coefficients. Some possible explanations include:

    (a) The assumption on the part of management that automation
         negates the need to pay attention to wastes.

    (b) Single use of cleaning compounds in large volumes.

    (c) Complexities and increased product loss in filling and con-
         veying operations.

    (d) Increased use of lubricants on cases, stacks and conveyors
         with automated materials handling.

    (e) Increased amount of product that can be left in lines (a
         1 1/2 million pound milk plant may have lines with
         a capacity up to 8 to 10,000 pounds of product).

    (f) Increased concentration of waste with more efficient water
         usage.

As dairy food plants become larger, they have a greater need to be located in an area that provides for optimum access to major highways. The tendency is to locate these large plants near interstate highways in suburban areas, frequently in small to medium-sized cities.

The trend is for dairy plants to service a larger and larger area, trucking in raw milk from considerable distances and hauling out packaged products in semitrailers as many as 500 miles from the processing plant. The independence of the plant location from closeness to the market has resulted in the trend to locate new large plants in suburban areas, some distance from any major waste treatment facility. Frequently, little attention has been given to the problem of waste disposal as a key criterion in the location. A number of new plants, discharging up to 8,000 lbs. of BOD/day, have been located in suburban areas or in cities of under 50,000 population. Where such plants utilize the municipal waste treatment facility they can become the major contributor to the BOD load of the municipal system.

The trend towards larger, centralized dairy foods plants generating 2,000 to 10,000 lbs. of BOD/day, will frequently place an additional burden on municipal treatment plants that may already be marginal in their operation. The average milk plant of 1980 (250,000 lbs. of milk/day) can be predicted to have wasteloads with a population equivalent of about 55,000 unless special effort is made to pretreat and/or markedly reduce these wastes.

REFERENCES

(1) *Bureau of Labor Statistics Bulletin 1737.* 1972.

(2) Harper, W.J., J.L. Blaisdell and J. Grosshopf. *Dairy Food Plant Wastes and Waste Treatment Practices.* Environmental Protection Agency, 12060 EGU 03/71. 1971.

(3) Federal Water Pollution Control Administration. 1967. "The Cost of Clean Water." Vol. III. *Industrial Waste Profile No. 9 – Dairies.* Fed. Wat. Poll. Cont. Admin., Publ. No. I.W.P.-9, U.S. Govt. Printing Office, Washington, D.C.

(4) Anon. "An Economic Analysis of Whey Utilization and Disposal in Wisconsin." *Ag. Econ.* 44. Univ. of Wisconsin, Madison. 1965.

(5) Anon. *Proceedings Whey Utilization Conference USDA-ARS-73-69.* 1970.

(6) Arbuckle, W.S. and L.F. Blanton. *The Development, Evaluation and Content of a Pilot Program in Dairy Utilization (Dairy Waste Disposal and Whey Processing).* Cooperative Extension Service. Univ. of Maryland.

(7) Jones, V.A. "Is Water 'Drowning' Profits in your Plant?" *The Milk Products Journal.* 50:10-11. 1959.

(8) Zall, R.R. and W.K. Jordan. "Monitoring Milk Plant Waste Effluent – A New Tool for Plant Management," *Journal of Milk and Food Technology.* 32. pp. 197-202. 1969.

(9) Watson, C.W. "Dairy Products." Chapter 5 in *Industrial Wastewater Control.* Academic Press, N.Y., N.Y. Edited by C.F. Gurnham. 1965.

(10) Watson, C.W. "Your Dairy Waste Line is Showing." *Milk Industry Foundation Convention Proceedings.* pp. 12-24. 1961.

(11) Anon. "An Industrial Waste Guide to the Milk Processing Industry." *Public Health Service Publication No. 298.* 1963.

(12) Nemerow, N.S. *Theories and Practices of Industrial Waste Treatment.* Addison-Wesley Pub. Co., Inc., Reading, Mass. 1963.

(13) Watson, K.S. "The Treatment of Dairy Plant Wastes." Prepared for the Environmental Protection Agency Technology Transfer Seminar, Madison, Wisconsin (Mar. 20-21, 1973).

(14) Carawan, R.E., V.A. Jones and A.P. Hansen. "Water and Wastewater Management in Dairy Processing." *Report PB 220,704.* Springfield, Va. Natl. Tech. Info. Service (Dec. 1972).

(15) Harper, W.J. and J.L. Blaisdell. "State of the Art of Dairy Food Plant Wastes and Waste Treatment." Second National Symposium on Food Processing Wastes. March 23-26, 1971.

(16) Trebler, H.A. and H.J. Harding. "Dairy Waste Disposal." *Chem. Eng. Prog.*, 43:(5)255.

(17) Porges, N. "Practical Application of Laboratory Data to Dairy Waste Treatment." *Fd. Technol.*, 12:(2)78-80. 1958.

(17a) Wallgren, K. H. Leesment and F. Magnusson. "Investigations of Irrigation with Dairy Waste Water." *Meddn. svenska Mejeriern.* Riksforen., 85:20. 1967.

(18) Sarkka, M., J. Nordlund, M. Pankakoski and M. Heikonen. "Water Pollution by Finnish Dairies." *18th Int. Dairy Congr.*, I-E, A. 1.2 11. 1970.

(19) Dairy Effluents Sub-Committee on the Milk and Milk Products Technical Advisory Committee. *Dairy Effluents.* Department of Agriculture and Fisheries for Scotland. 1969.

(20) Trebler, H.A. and H.G. Harding. "United States Trends in Disposal of Dairy Waste Waters." *12th Int. Dairy Congr.*, 3:Sect. 3, 688-697. 1949.

(21) Trebler, H.A. and H.G. Harding. "Milk Waste Problem, Review." *Sewage Ind. Wastes*, 20:(3)594. 1948.

(22) Dairy Industry Committee, Subcommittee on Dairy Waste. "Dairy Waste Subcommittee Meets in Upstate New York." *Milk Prod. J.*, 48:(6)82.

(23) Eldridge, E.F. "Waste Prevention in the Dairy Industry." *Proc. Conf. Ind. Wastes*, Eng. Exp. Sta. Washington Univ., 57-59. 1949.

(24) Fisher, W.J. "Treatment and Disposal of Dairy Waste Waters: A Review." *Dairy Sci. Abstr.* 30:(11)567-577. 1968.

(25) Kountz, R.R. "Dairy Waste Treatment." *J. Milk Fd. Technol.*, 18:243-245. 1955.

(26) Wooding, N.H., Jr. "Waste Management, Treatment and Disposal for the Food Processing Industry." Special Circular 113. Penn State University, University Park, Pa.

(27) Zack, S.I. "Treatment and Disposal of Milk Wastes." *Sewage Ind. Wastes*, 25:177-187. 1953.

(28) Baustian, H. "Utilization of Whey in Cheese Factories." *Molk.-u. Kas.-Ztg.*, 12:(1)9-10. 1961.

(29) Groves, F.W. and T.F. Graff. "An Economic Analysis of Whey Utilization and Disposal in Wisconsin." *Univ. of Wis. College of Agr. and Agr. Econ. Bull. 44.* 1965.

(29a) Groves, F.W. "An Economic Analysis of Whey Utilization." *Bulletin No. 48*, University of Wisconsin (July 1972).

(30) Hanrahen, F.P. and B.H. Webb. "Spray Drying Cottage Cheese Whey." *J. Dairy Sci.*, 44:1171. 1961.

(31) Jasewicz, L. and N. Porges. "Whey Preservation by Hydrogen Peroxide." *J. Dairy Sci.*, 42:1119. 1959.

(32) Mickelsen, R. "Dairy Wastes and the Pollution Problem." *Mfg. Milk Prod. J.*, 58:(3)8. 1967.

(33) Oborn, J. "A Review of Methods Available for Whey Utilization in Australia." *Aust. J. Dairy Technol.*, 23:131. 1968.

(34) Renwick, R.S. "Utilization of Milk By-Products: Skim Milk." *J. Soc. Dairy Ind.*, 16:50. 1963.

(35) Sprague, G.W. "What Has Happened to Whey?" *Mfg. Milk Prod. J.*, 45:(11)26. 1954.

(36) Stringer, W.E. "Whey Once Discarded Now Provide Useful Products." *Fd. Inds.*, 21:892-895. 1949.

(37) Wasserman, A.E. and N. Porges. "Whey Utilization: Summary of Laboratory Investigations in Yeast Propagation." *Proc. 14th Ind. Waste Conf.*, Purdue Univ., 535-546. 1959.

(38) Werner, H. "Whey as an Effluent Problem." *Maelkeritidende*, 83:(19)431-438. 1970.

(39) Whittier, W.O. and B.H. Webb. *By-Products from Milk*. New York: Reinhold Publishing Company, p. 33. 1950.

(40) Anon. "Spray Drying Adapted for High-Acid Whey." *Chem. Eng. News*, 39:(25)66. 1961.

(41) Calvert, H.T. "Milk Factory Effluents." *Rep. Dir. Wat. Pollut. Res.*, Lond., 16-21. 1939.

(42) Houran, G.A. "Practical Uses of Whey." *17th Ann. Dairy Eng. Conf.*, Mich. St. Univ., 69-74. 1969.

(43) Horton, B.S., W. Eykamp, R.L. Goldsmith and S. Hossain. "Membrane Separation Processes for the Abatement of Pollution from Whey." *18th Int. Dairy Congr.*, I-E, A. 4. 10, 442. 1970.

(44) McDonough, F.E. "Whey Concentration by Reverse Osmosis." *Fd. Eng.*, 40:(3)124-127. 1968.

(45) Wyeth Laboratories. "Wyeth Uses Electrodialysis to Desalt Whey." *Chem. Eng. News*, 40:(41)44-45. 1962.

(46) Hare, J.H. and B.E. Baker. "The Precipitation of Whey Proteins Using Waste Sulphite Liquor." *Can. J. Technol.*, 29:332-336. 1951.

(47) Amundson, C.H. "A 1967 Solution for Italian Cheese Whey Disposal Increasing the Protein Content of Whey." 4th Annual Marschall Invitational Italian Cheese Seminar, Madison, Wisconsin. 1967.

(48) Anon. "Yeast Producer Thrives on B_{12}." *Chem. Eng.*, 59:244 and 246. 1952.

(49) Enebo, L., H. Lundin and K. Myrback. "Yeast from Whey." *Svenska Kem. Tidskr.*, 53:96. 1941.

(50) Atkin, C., L.D. Witter and Z.J. Ordal. "Continuous Propagation of *Trichosporon cutaneum* in Cheese Whey." *Appl. Microbiol.*, 15:(6)1339-1344. 1967.

(51) Bridgewater, N.P. and A. Taylor. "Effluent Disposal in the Food Industry." *Fd. Mf.*, 22:127. 1947.

(52) Rugosa, M., H.A. Browne and E.O. Whittier. "Ethyl Alcohol from Whey." *J. Dairy Sci.*, 30:263. 1947.

(53) Svoboda, M. "Studies on the Purification of Dairy Effluent by Fermentation." *Zpravy Vyzkumneho ustavu pro mleko a vejce*, Praha, (3/4)17-22. 1955.

(54) Carrick, C.W. "Use of Industrial Wastes in Poultry Feedings." *Proc. 6th Ind. Waste Conf.*, Purdue Univ., 130-134. 1951.

(55) Dias, F.F. and J.V. Bhat. "Microbial Exology of Activated Sludge. I. Dominant Bacteria. II. Bacteriophages, *Bdellovibrio*, Coliforms and Other Organisms." *Appl. Microbiol.*, 12:412. 1964.

(56) Adamse, A.D. "Bacteriological Studies on Dairy Waste Activated Sludge." *Meded. LandbHogesch.* Wageningen, 66:(6)1-79. 1967.

(57) Pipes, W.O. *An Atlas of Activated Sludge*. Federal Water Pollution Control Administration. U.S. Department of Interior. 1968.

(58) Boyle, W.C. and L.B. Polkowski. "Alternate Methods of Treating or Pretreating Dairy Plant Wastes" in paper before EPA Technology Transfer Seminar, Madison, Wisconsin (March 20-21, 1973).

(59) A.T. Kearney and Co., Inc. "Study of Wastes and Effluent Requirements of the Dairy Industry." Water Quality Office, Environmental Protection Agency, Cont. No. 68-01-0023, July, 1971.

(60) Culp, R.L. and G.L. Culp. *Advanced Wastewater Treatment*. New York: Van Nostrand Reinhold Co. 1971.

(61) Aqua-Chem. Inc. for Office of Saline Water. "Industrial By-Product Recovery by Desalination Techniques. *R & D Progress Report No. 581* (Oct. 1970).

(62) Hickman, P.T. "The Benefits of Joint Treatment Approach with the City," paper presented at EPA Technology Transfer Seminar, Madison, Wisconsin (Mar. 20-21, 1973).

(63) Anon. *Federal Register.* Vol. 35, No. 128. July 2, 1970.

(64) Anon. "Industrial Wastewater Surcharges." *Water Research for Action Report No. 1*, Water Resources Research Institute, North Carolina State University. 1972.

(65) Zall, R.R. "Monitoring Waste Discharge: A New Tool for Plant Management." Dissertation, Cornell University. 1968.

(66) Boyle, W.C. and L.B. Polkowski. "Treatment of Cheese Processing Wastewaters in Aerated Lagoons." *Proc. Third National Symposium on Food Processing Wastes*, New Orleans, La., March, 1972.

(67) Birks, C.W. and R.J. Hynek. "Treatment of Cheese Processing Wastes." *Proc. 26th Purdue Industrial Waste Conference*, May 4-6, 1971.

(68) Homel, J.A., Foth & Van Dyke, Green Bay, Wisconsin. *Engineering Design and Performance Analysis of Eiler Cheese Plant Waste Water Treatment Facilities.*

(69) Foy, R., Carl C. Crane & Assoc. Inc., Madison Wisconsin. *Engineering Design of Afolkey Coop. Cheese Co. Wastewater Treatment Facilities.*

NOTICE

Nothing contained in this Review shall be construed to constitute a permission or recommendation to practice any invention covered by any patent without a license from the patent owners. Further, neither the author nor the publisher assumes any liability with respect to the use of, or for damages resulting from the use of, any information, apparatus, method or process described in this Review.

MARSTON SCIENCE LIBRARY

Date Due

Due	Returned	Due	Returned

Form #0067